D1162910

Courtship: an ethological study

by Margaret Bastock

What is courtship? Why do animals court? Why is animal courtship often elaborate and seemingly fantastic? To what extent are techniques of courtship learned?

This concise but thorough study of courtship behavior in fish, birds, and arthropods is the first rigorous examination of the evolutionary origins and mechanisms of courtship and its contribution to biological success. Demonstrating the fruitfulness of an empirically based, inductive approach to understanding courtship, the book also explains clearly how principles of modern evolutionary theory can be successfully employed in studying behavior.

The author describes many observations and experiments that have not previously appeared outside specialized journals and brings an abundance of simple yet accurate examples of animal behavior to bear on explanations of ethological concepts and evolutionary theory. No attempt is made to skim over the gaps of knowledge apparent in the study of behavior evolution; rather, the author discusses the limitations and difficulties of different approaches, critically reviews the deductions that can be and have been made from them, and tries to present enough evidence on controversial points for the reader himself to judge the validity of specific arguments.

Courtship

Courtship
An Ethological Study

Margaret Bastock

ALDINE PUBLISHING COMPANY
CHICAGO

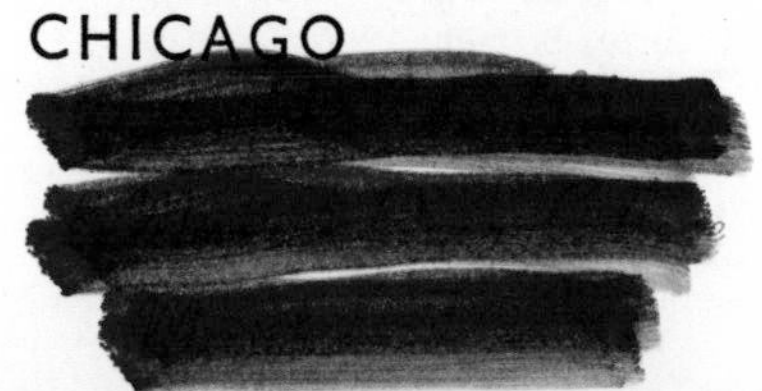

First published in the United States of America by
ALDINE Publishing Company
320 West Adams Street
Chicago, Illinois 60606

Heinemann Educational Books Ltd, London

Library of Congress Catalog Card Number 67–27325
Printed in Great Britain

Preface

A number of books have described courtship, often in great detail. Some have also speculated on its function and evolution but few have discussed its importance in the study of animal behaviour as a whole.

Courtship has many features which commend it to behaviour workers. Field studies gain from its exhibitionist qualities while genetical and evolutionary investigations are aided by the fact that the displays tend to show distinctive differences between species while remaining very uniform within each population. Two further features raise problems of origin and causation: courtship is an unexpected activity occurring when one would often expect only mating behaviour, and courtship has remarkable variety of form, often bordering on the fantastic. A new and very successful approach to these old problems comes from the work of modern ethologists and it is their arguments which I present in this book. Ethologists stress the observation of behaviour under natural conditions and deduce facts about its causation from the circumstances in which it occurs and from the behaviour which accompanies it.

There are features too, which favour the analytical study of courtship. Many external features (like the presence of nests or eggs or other animals) can modify the frequency or form of courtship displays and it is easy to alter these without interfering too much with the animal. And experiments on the physiological basis of courtship are suggested by facts already known about the physiology of reproduction.

Any study of one particular aspect of a given activity (its physiology, its evolution, its development) is necessarily incomplete in itself. In this book I have tried to build up a picture of courtship as part of the animal's whole behavioural organization, taking account of its history as well as its present form. I have considered the role it plays, how it operates and how it came to do so. I have not attempted to describe courtship in every animal group nor have I reviewed all relevant studies. Instead I have selected those investigations which appear to be most complete and where a coherent picture emerges for at least one group of animals. Inevitably different types of study have centred upon different groups. Some of the best physiological work has been done on birds, neurophysiological studies have concentrated on mammals while genetical analyses are well illustrated from insects. Hence I have tried to indicate the type of process involved rather than particularize: at the same time I have given examples from a variety of animal groups to show that the principles demonstrated are (often surprisingly) of wide application.

The chief omissions include much interesting (although scattered) work on descriptions of mammalian displays and on insect physiology. I regret these but a short book cannot be comprehensive. Both insects and mammals are well represented in various sections of the book. In particular, theories regarding courtship as an example of 'conflict' behaviour are of especial relevance to higher animals and I hope that psychologists as well as biologists will find something to interest them here.

My thanks are due especially to S. A. Barnett, my editor, for his very constructive criticisms, and also to Aubrey Manning for the considerable time he has spent reading, criticizing and discussing. These two, in particular, have both encouraged me and led me to think again on many points. But I must also acknowledge my debt to the thinking and teaching of Professor N. Tinbergen at Oxford and to many friends and colleagues for discussions and encouragement.

M. B.

Contents

Introduction

There is usually no difficulty in recognizing courtship. The males of many species display to females immediately before mating; they dance, posture, call and otherwise behave peculiarly and conspicuously. Often females join in, sometimes displaying in the same way, sometimes reacting to male displays with different ones of their own.

In common usage display means conspicuous behaviour designed to attract attention and show something off. This implies communication between displayer and audience, and in scientific usage this aspect is stressed. A cormorant looks conspicuous when it spreads its wings to dry them, but unless it can be shown that this behaviour has some effect upon another cormorant, it is not called a display. Hence the term 'courtship display' implies that a male is influencing the behaviour of a female, or that a female is influencing a male.

However, the strict definition of courtship is made difficult by the fact that some species have several distinct displays which occur at different times during the period before mating. Many birds pair early in the year (February or March) although they do not mate until April or May. Most have distinct displays immediately before mating, and it is possible to single these out and label them 'courtship'. But, in addition, they may have separate displays at pairing time as well as nest-site displays, nest-building displays and greeting ceremonies, among others, during the intervening engagement period. These often share elements in common. A male green heron adopts a particular posture (called the stretch) both in the nest-site display and (sometimes) in the display before coitus. Further, where several distinct signals are involved, any given display may include one or more of them. The kittiwake greeting ceremony is also a nest-site display; so is the stickleback sexual dance. Indeed, whatever their special function, all these various displays are probably sexually stimulating. Hence it is impossible to draw any general distinction between them either on the basis of form or of function. D. Morris has proposed a definition which includes them all: 'courtship is the heterosexual reproductive communication system leading up to the consummatory sexual act' [195]. This is the definition adopted in this book.

The question of function

Why do animals court? This is perhaps the first question which springs to mind when one considers courtship. Sexual reproduction is a usual feature of animal life, and it commonly involves the close association of a single pair of animals. There may be internal insemination, external fertilization (the male fertilizing the eggs as soon as they are laid), or simultaneous spawning (male and female shedding their sexual products simultaneously, close together). With all these methods, there seems no reason why, once they have met, sexually mature males and females should not mate forthwith. And this is just what does occur in some animals. Although courtship is the general rule for mammals and birds, it is sporadic among reptiles, amphibians, fish and invertebrates. A male house-fly *Musca* leaps on to a female's back without preliminary, while a male fruit-fly *Drosophila* performs an elaborate dance before attempting to do so.

The question has, in fact, no single answer. Many advantages of courtship will be discussed in this book, but their importance varies with circumstances. Others may have been overlooked, for it is not yet possible to state all the necessary conditions for courtship.

The question of complexity

Why is animal courtship often so elaborate and fantastic? When one considers the bower birds (Ptinolorhynchidae) [179] building and decorating elaborate nests to entice females, or terns *Sternus* [11, 73] performing intricate display flights with their females, it is natural to ask whether such extraordinary preliminaries are really essential for mating.

This question could be reworded: 'what determines the form of courtship?' and this, together with the first question, 'why do animals court?' is an ambiguous one. Both can be answered in two very different ways – namely, in terms of evolutionary significance or in terms of immediate mechanism. In an evolutionary explanation the functions of courtship are considered as possible reasons for its evolution. If a courting animal is 'fitter' than a non-courter, in the sense that it is better able to survive and reproduce, then its descendants (in so far as they inherit these qualities) will prosper at the expense of others; and more and more animals will come to be courters in subsequent generations. Moreover, if one courter has a particular form of display which makes it fitter than the others, then that display too will spread in the population. Hence it is a legitimate evolutionary explanation of a male's courtship display to say that it enhances his prospects of obtaining a mate without in any way diminishing his chances of survival.

An explanation in terms of mechanism relates to factors within and without the animal which promote courtship. We might find that a male courts because sex hormones, circulating in the blood, prime certain nervous mechanisms to respond to stimuli provided by the female. The form of the response would then be understood in terms of the nervous organization underlying it. This kind of explanation is usually the only one open to experimental investigation. Where evolutionary advantages are concerned there are difficulties, because we can never reproduce the original situation. In a population of courters, we can test the advantages of normal courters over abnormal ones or over individuals in which the 'instruments of courtship' (like peacock's tails) have been removed or modified. But these are not necessarily comparable to the advantages possessed by rare courting animals in a population of non-courters. The variations concerned may be different in nature, with quite different effects upon total fitness. The populations and their circumstances will be different too.

Nevertheless it is stultifying to ignore evolutionary explanations. In fact courtship illustrates well how the two types of explanation are related. The form of a display is rarely completely 'explained' in terms of the advantages it confers. There must be many ways for a male to attract, arrest or excite a female. Anatomical and physiological mechanisms, already present, may limit the possibilities; but these too are products of evolution. Form, function and mechanism are as much related to their evolutionary history as they are to one another.

Natural selection and sexual selection

The evolutionary explanation of courtship (and of sexual adornments) has, in the past, aroused considerable controversy concerning the nature of the advantages conferred. It is usually accepted that these characters are concerned with reproductive fitness rather than with survival. As such they are no less important, since a sterile animal, however long it lives, cannot influence subsequent generations. But reproductive fitness depends upon many factors, among them obtaining a mate, as well as producing and rearing healthy offspring. Charles Darwin singled out the winning of a mate as an exceptionally competitive process in which displays give one male advantage over another [77]. In other words, he believed that displays and adornments are advertising gimmicks which attract females to one particular male, just as trading stamps may attract customers to a particular shop without regard to any real value. Indeed he argued that they might even conflict with the demands of other aspects of fitness. A peacock's tail may help him to win more females, but it is useless as regards the rearing of young and even a hindrance if he needs to flee from predators.

However, J. S. Huxley and others have argued that mating is not always a simple process, as Darwin supposed, and that courtship can be important in facilitating it [125, 187]. Sometimes it helps male and female to find one another, to indicate to one another suitable breeding sites and to synchronize physiological processes so that ova and sperm are ready for fertilization together. And sometimes it may help break down other barriers which prevent the mates from coming together.

Undoubtedly both these types of advantage exist; their relative importance in evolution no doubt varies with the species. The evolution of courtship, like that of any other character, can be initiated only by some changed circumstance, shifting the balance of advantages. The nature of the advantage depends upon the nature of the change. An increase in population size might increase male competition and promote sexual selection; a decrease might favour signals helping mates to meet. And displays, once evolved, may acquire new advantages in new contingencies, so that we may now see many present-day functions but can only guess their significance in evolutionary history.

Courtship and conflict

Consideration of the origins of courtship raises a further question What did it evolve from? We may ask how some animals could have come to perform any particular action sufficiently often before mating for its advantages to make themselves felt. The answer is most likely to come from observations of animals in their natural environments. In particular, Tinbergen [270, 271] saw that there might be some significance in the fact that courting animals hold territories at breeding time, and that territorial and mating behaviour often coincide.

Breeding territories are common among mammals, birds and reptiles, sporadic in amphibians and fish, and rare among invertebrates. Usually the males of a species establish themselves, early in the breeding season, each in a separate locality which they defend from other males of the same species. They proclaim their occupancy by displaying themselves conspicuously therein, by patrolling the boundaries and attacking intruders. Females rarely set up territories, but visit the males in theirs. Later, pairing, courtship, mating and breeding may occur within the territory, since male and female often remain there together until their young are reared.

Territorial behaviour probably has many advantages for a breeding animal. It provides a base from which the male can advertise himself to females and where, later, the pair can find one another. It provides a protected place where courtship, mating and nest-building can occur, free from interference (although the degree of such immunity varies with the size of the territory). And spacing out also provides both some protection from predators and also (sometimes) a private food supply close at hand [58, 115, 121, 122, 141, 199, 272].

However, territorial behaviour can complicate the processes of pairing and mating by introducing an element of hostility. A territorial male's first reaction to a female visitor is usually aggressive, since she is an intruder and possesses at least some of the characteristics of males of her species. Hence territorial males often seem ambivalent in their approaches to females: attacks may alternate with, or interfere with, sexual approaches. In other words, these early encounters occur in a situation which invokes two opposed kinds of behaviour. Situations of this type are often called conflict situations, and in them many animals (sometimes including man) behave in characteristic ways. They are restless and disturbed; they change rapidly from one activity to another; their movements may be incomplete or distorted; and they do apparently irrelevant things. Such behaviour is collectively called conflict behaviour, and Lorenz and Tinbergen see it in a clue to the origins of courtship. They suggest that similar conflicts existed in ancestral, as in present-day, animals and that in some circumstances the conflict behaviour they performed helped improve their mating success and hence evolved into displays.

This is an important suggestion, although it is not the whole of the story. Almost all territorial animals court, but in those with long engagement periods these proposals apply only to pairing displays. Hostility between the mates quickly dies down. Moreover, there are non-territorial animals which also display. We have to ask whether other conflicts exist in these cases.

Problems of inheritance

It is quite plausible that one bird might win a mate more easily than another because he preens rather than bathes when he encounters a female. Preening might display his attractive colours better. But will his offspring inherit this useful habit? Certainly they may do so. They will if his preening habit depends upon a heritable difference in physiology or nervous organization (perhaps causing him to preen more often in general). On the other hand, his changed behaviour may depend upon changed circumstances; he may preen rather than bathe because he inhabits a drier territory with less water for bathing. Then his offspring will repeat his behaviour only under similar circumstances. However, there is increasing evidence that genetical factors can affect the circumstances of an animal's life as well as his reactions to them. Thus the choice of a drier rather than a damper territory could be the result of an inherited physiological difference. In this way offspring may inherit a new habit indirectly, by tending to provide the conditions which favour it.

Problems of mechanism

Courtship is often called an 'instinctive activity', but this term is difficult to define. However, we may say that the courtship of each species is very uniform and occurs normally in response to certain special stimuli. It is usually complete and recognizable even at the first performance. A nervous organization must develop within the animal, before sexual maturity, which determines both the form of courtship movements and their link with significant stimuli. We may ask what parts of the nervous system are involved, and how they operate. We may ask how far form and sequence in courtship depend upon internal organization, and how far they are related to the circumstances of courtship. And we may ask how such mechanisms develop: how inherited and environmental factors interact to produce them.

Courtship is not evoked, in any animal, with mechanical consistency. A male, encountering a female, may on one occasion court her intensely, on another, ignore her. Variations in responsiveness such as these are typical of 'instinctive' activities and are often called changes in *motivation*. However, the use of this term suggests that we are dealing with a single underlying physiological process, and this is unlikely to be the case. Sexual responsiveness can be measured in various ways: the immediacy of the response, its duration, its intensity, its sensitivity (judged by the strength of stimulus required to elicit it). There is not always a correlation between them. Moreover, many factors affect sexual responsiveness. It may decline temporarily just after coitus, or after prolonged unsuccessful courtship;

it may drop sharply if an enemy appears; it may fall gradually if the animal becomes starved or ill; it may disappear entirely during the winter months. Some of these changes are sudden, some gradual; some temporary, some prolonged. Probably, different factors modify the reactivity of the mechanism in quite different ways. If the term 'motivational changes' is used, it must not be allowed to obscure the need to investigate the physiological effect of each of them separately.

These are only a few of the problems raised by the study of animal courtship. This book gives an account of some scientific attempts to elucidate them.

Part One: Some Descriptions

1: Fish

Fish vary greatly in their reproductive habits. Some are territorial, most are not; some are social, others are not. Their mating habits vary too. Most are polygamous, each male fertilizing (externally) the eggs of many females. But some are monogamous, some have internal fertilization and a few are viviparous. Parental behaviour ranges from the guarding of eggs and young by one or both parents, through the abandonment of eggs in carefully prepared places, to the simple abandonment of eggs anywhere in the sea (figure 1). The following descriptions will show to what extent courtship varies with these factors.

Some account of the behaviour associated with fighting inevitably enters into these accounts; it is therefore necessary to define the terms which are employed as they are often loosely used with subjective and anthropomorphic implications. In this book *aggression* means attack behaviour and *aggressive reactions* are the postures and movements associated with attacks. An animal is said to be *aggressive* when it shows a high tendency to attack. Escape and withdrawal activities are unambiguous but the term *fear reaction* is used (without subjective implications) to imply the adoption of postures and movements associated with withdrawal; similarly an animal is said to be *frightened* if it shows fear reactions and is likely to escape. The whole complex of behaviour patterns associated with fighting, including both aggression and escape, is sometimes called *agonistic behaviour.* Two displays are common in agonistic situations. Both are best defined in terms of their function. *Threat* displays tend to cause withdrawal on the part of the adversary; *appeasement* or *submissive displays* tend to reduce attacks. Each is given in special circumstances, threats usually by aggressive animals and appeasement

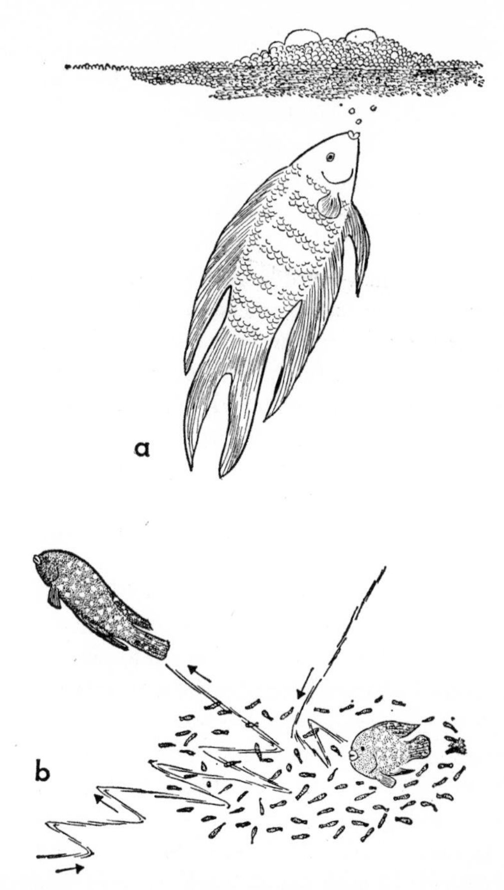

Figure 1. Two types of parental behaviour in fishes.
(*a*) The paradise fish *Macropodus opercularis* builds a bubble nest, blows the fertilized eggs into it, and abandons them.
(after Peters [209a], from Portman [215].)
(*b*) Male and female jewel fish *Hemichromis bimaculatus* take turns to guard the young in special pits. The female (right) has just relieved the male (left), who is departing swiftly so that the young do not follow.)
(After Baerends and Baerends [9].)

displays by frightened animals. They will be discussed more fully later.

Four examples of fish courtship are given. The stickleback is a polygamous, territorial fish with external fertilization and parental care; the jewel fish is similar but monogamous; the whitefish is a non-territorial simultaneous spawner which abandons its eggs; the guppy is a non-territorial, viviparous, social fish with internal fertilization.

The three-spined stickleback

The courtship of the stickleback *Gasterosteus aculeatus* has been described mainly by Tinbergen and his pupils [208, 269, 271]. Outside the breeding season, this fish tends to live in a school, that is to say in a peaceful communal group which has no leader and in which there is no fighting. The most striking feature of a school is that the individuals composing it usually perform the same activities at the same time: for example, they swim around together or feed together. In early spring the whole community migrates to breeding grounds in freshwater streams or ponds. The males develop a nuptial colouring (featuring a bright red belly and bright blue eyes), leave the school and set up territories in an area of sandy bottom containing water plants. Like most fish territories, these are small and closely grouped.

The early stages of territorial behaviour are mostly concerned with establishing the territory, and nest-building. The nest is of plant material laid in a covered pit dug in the sandy bottom. The territory extends outwards and upwards from this centre, but often ends some distance below the surface, so that fish can swim freely above it. In fact, this is often the only safe area in a cluster of territories, and females usually approach from here. Male intruders, by contrast, approach from their own territories at a lower level.

When the nest is nearly complete, a male begins increasingly to advertise for females, swimming close to the surface in a curious jerky fashion. Nevertheless, when one does appear and enters his territory, he is as likely to attack as to court her. However, ripe females are often remarkably bold intruders, in contrast to males, who are timid off their own ground. Females do not flee, even when butted and bitten; instead they adopt the so-called 'appeasement posture', hanging obliquely upwards in the water. This posture is also adopted by male rivals defeated in a fight. It nearly always inhibits attack, at least momentarily, and offers a chance of escape for males and a respite for females. The posture also displays the belly: in females this is silver (not red) and swollen with eggs. Hence female intruders differ from males both in behaviour and appearance, and their special characteristics appear to damp-down attacks and to stimulate courtship. This at first alternates with attacks, but may eventually dominate.

The most conspicuous feature of courtship is the *zig-zag* dance, whereby a male, swimming jerkily, alternates leaping towards the female with a sideways jump away from her. He may circle round her, but his sideways jumps are always roughly in the direction of the nest. From time to time one of these turns into a fast excursion to the nest, known as *leading*.

Courtship can proceed no further unless the female follows. (If she does not the male returns to his zig-zagging.) At the nest, he *shows* his nest entrance by lying on his side and thrusting his snout into it. The female must respond by pushing her way inside before the male can proceed to *quiver*, with his head vibrating against her tail. Then she usually spawns, and wriggles out the other side; the male follows her through the nest and fertilizes the eggs.

Very soon the female is chased away; the male repairs his nest and advertises for fresh females. Subsequent courtships are similar, although there now tends to be less overt aggression. Each male collects the eggs of about five females in his nest, if he can. Then he guards them and fans them until they hatch. Females play no part in parental care.

This chain of events in courtship ensures that the final spawning occurs in a properly completed nest between fully mature mates. A female who is not quite ripe will fail to respond at some point, and although the male may continue to court her for a while, he will eventually chase her away. Alternatively, an 'unready' male, particularly one whose nest is not quite complete, is likely to be too aggressive and frighten females away. The stimulating qualities of courtship appear to be effective only for fish already on the point of response. Nevertheless they are essential. In particular, quivering induces spawning in a female, as can be shown by imitating with a glass rod the male's prodding on her back. And some part of courtship seems to diminish the aggression of the male so that he allows a female near his nest; ripe females, which happen to arrive at the nest by chance, are almost invariably chased away.

The jewel fish

The jewel fish *Hemichromis bimaculatus* is a tropical member of the family Cichlidae. It has been studied in great detail by Baerends and Baerends-van Roon [9]. It resembles the stickleback in having out-of-season schooling behaviour and in being territorial at breeding time. It also lays eggs and has external fertilization. It differs from the stickleback in three important features of breeding behaviour, each of which has some effect upon courtship. First, female jewel fish hold territories before mating and therefore tend to be aggressive like the males; secondly, there is no nest, and male and female choose a spawning site together; thirdly, a single pair stays together for the whole season, and tend the young.

The initial stage of courtship is greatly affected by female territoriality. Females of only a few species of fish behave in this way, and little is known of its significance. The territories are held briefly at the beginning of the mating season, and are unlikely to have any importance as feeding areas. Nevertheless the females defend them vigorously and have bright nuptial colours, almost, but not quite, identical to those of the male; these they display in threat. This suggests that striking reproductive colours are as much associated with threat behaviour as with attracting a mate.

Female territories are quite suddenly abandoned, probably at the time of the ripening of the ovaries, and their owners begin to visit males in neighbouring territories. Visiting females must choose territories belonging to fish of the right species and the right sex. As in the stickleback, bright red markings form specific cues (other cichlids are green or blue); and female jewel fish are attracted to red males, in preference even to pink, when offered a choice. However, this does not prevent confusion between the sexes: inexperienced females do sometimes enter the territories of other females and even court them. Experienced fish rarely make this mistake, and it is believed that distinctive details of movement, colour or form are learned [202].

The reaction of the male jewel fish to an intruder is much the same as that of the stickleback; he attacks or displays aggressively. Sometimes he charges and butts in the head or abdomen, more commonly he makes a frontal attack which usually leads to mouth fighting (figure 2), each fish gripping the jaw of the other.

Female jewel fish, however, differ from female sticklebacks in that they fight back. They give threatening displays like the males and defend themselves from their attacks. However, female jewel

Figure 2. Mouth fighting in the jewel fish *Hemichromis bimaculatus*. (After Baerends and Baerends [9].)

fish are like sticklebacks in being unusually bold, in failing to flee, and in possessing an appeasement posture which indeed is very similar to that of the stickleback (figure 3*a*). This display is given increasingly as time goes on, when females begin more and more to dodge attacks rather than retaliate. Eventually, in this way, aggression between the pair dies down.

The sexual stage of courtship is mainly concerned with the selection and display of a suitable spawning site. However, courtship is by no means continuous; there may be periods of rest or feeding. It is tempting to relate this more leisurely affair to the fact that the male jewel fish, unlike sticklebacks, is in no hurry to get rid of one female in order to hunt for more. The displays are very similar for both male and female. Either fish may initiate any of them, and each copies the other, displaying alternately. They occur in three phases. In the first, one fish selects a suitable stone or shell for spawning and makes movements towards it called *jerking* and *quivering*. Jerking consists of pointing the body towards the stone and jerking the head once or twice, first to one side, then to the other. In quivering (figure 3*b*), the body is vertical over the chosen object and the whole body trembles. Occasionally digging is seen, when sand is carried

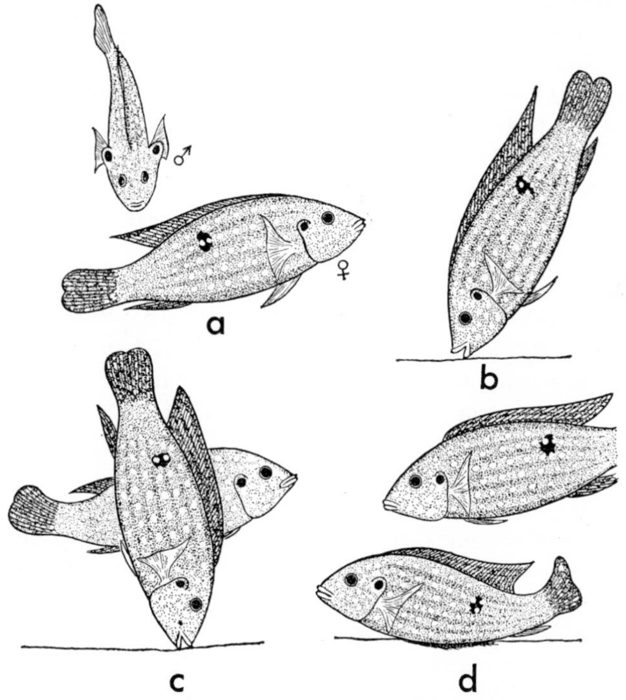

Figure 3. The courtship of the jewel fish, *Hemichromis bimaculatus.* (After Baerends and Baerends [9].)

(*a*) Threat display by male; appeasement by female; (*b*) quivering; (*c*) nipping; (*d*) skimming, leading to spawning and fertilizing.

away from the bottom to form a small pit. Such pits play no part in mating, however, but are used later for the care of young.

When both fishes have jerked and quivered for a while, one partner begins a new display called *nipping* (figure 3*c*). Plants or dirt are nipped off the stone and carried away. Thus the spawning site is cleaned, although the behaviour occurs whether cleaning is required or not. Finally the fish change to the third phase, of *skimming* (figure 3*d*), when they glide over the stone touching it with their abdomens. The tempo increases, and eventually the female lays a batch of eggs during one of her skims and the male, following, fertilizes them.

A female lays several batches of eggs and afterwards stays near them, guarding them and often driving the male away. Later the young are carried to small pits which have been dug meanwhile. Male and female then alternate in parental duties, one patrolling while the other guards, although the male does more patrolling and the female more guarding.

The main features distinguishing jewel fish from stickleback courtship are the prolonged early aggressive phase and the extraordinary parallelism of male and female behaviour throughout. The former could be accounted for by the fact that the female fights back and lengthy battles often develop. The prolongation of this phase is also useful in relation to the jewel fish's breeding habits; during it the mates become acquainted with details of each other's appearance, and learn not to fight one another while still attacking other members of the same species. This is essential if they are to live together for a long period on a territory which they must still defend. This feature may be compared with the lengthy pairing phases of many birds where, again, a single pair stays together on a territory to rear the young.

The parallelism of the displays is due, in part, to the fact that the choice of spawning site is shared. The male stickleback has to entice the female to an already-existing nest; each jewel fish has to signal acceptance of a suitable stone. The imitative behaviour is typical of that found in an 'out of season' school, and might be expected once hostility has died down; it can be compared with the flock behaviour seen during the engagement periods of some birds. Such tendencies have no time to develop during brief stickleback courtships.

In conclusion, stickleback and jewel fish courtship may be compared with that of another cichlid *Tilapia natalensis*, closely related to the jewel fish and also studied by Baerends and Baerends [9]. Here only the male holds a territory, and he digs a pit in it for spawning, before the appearance of the female. He is also polygamous;

each female leaves the territory immediately after mating, carrying the fertilized eggs in her mouth for oral 'incubation'. Some of these conditions are reminiscent of those in the stickleback. So is the courtship sequence illustrated in figure 4. There is no prolonged first stage: a male invites and leads a female (who is passive and does not fight in spite of some initial attacks); she follows his 'leading' display

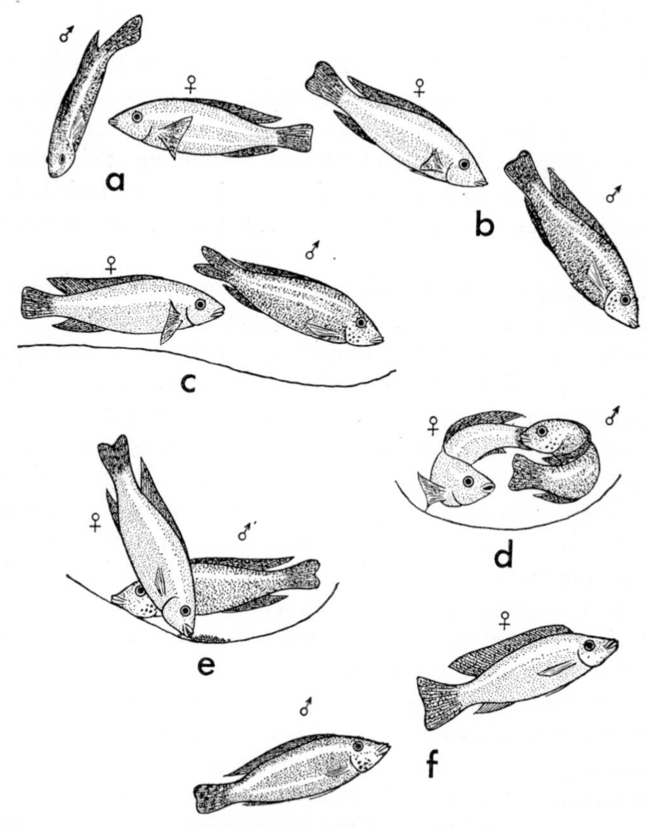

Figure 4. The courtship sequence in the cichlid fish *Tilapia natalensis* (After Baerends and Baerends [9].)

(*a*) Inviting; (*b*) leading; (*c*) tail wagging at pit; (*d*) circling; (*e*) male fertilizing, female snapping up eggs after spawning; (*f*) male chasing away spent female.

to the pit, where she spawns after mutual circling; the male fertilizes the eggs as soon as they are shed. In this case, then, a similar pattern of courtship appears under similar conditions in two distantly related fish, while two closely related ones, living in dissimilar conditions, have quite different patterns.

The whitefish

Whitefish *Coregonus lavaretus* have been studied by Fabricius [85]. They are marine, but migrate either into freshwater or into brackish coastal water for spawning. Normally they swim in dense schools and are exceptionally peaceful fish with no fighting at all either in or out of the breeding season. This is associated with their lack of territoriality, as is the fact that they have no conspicuous nuptial colours.

At breeding time the schools break up, although many of the fish stay in roughly the same area. Hence there is no difficulty in mate meeting mate, yet the fish are sufficiently scattered for uninterrupted courtship. The first sign of spawning tendency is a peculiar '*sailing*' display given by adults of both sexes. A fish feeding at the bottom may suddenly adopt a distinctive posture with all its fins extended. Then it swims in a slow, exaggerated manner, patrolling a small area of the bottom (two to three metres square) for perhaps several hours, although it makes no attempt to keep other fish away. Later it may become very lively, making dashes to the surface in the sailing posture and splashing a great deal. This perhaps functions as an advertisement, and attracts the attention of other fish, although they may not immediately approach.

Courtship and spawning usually occur at dusk or even in the dark. A male approaches another fish near the bottom and swims with it obliquely upwards, always against the stream. Each is in the sailing posture, their flanks touch, their movements are rapid and they perform spasmodic body undulations. In a successful courtship, simultaneous spawning occurs at or near the surface. Just before the orgasm the undulations of male and female become synchronized. Eggs fall away behind the pair in a cloud of sperm, and, because of the upstream course taken in courtship, many of them arrive back at the starting place. Some may be eaten on the way down by other fish, but the rest usually fall into crevices, since the preferred type of bottom is usually a mixture of sand and stone. Eggs are not tended in any way. After spawning the partners separate, and each perhaps immediately repeats the process with another fish nearby.

The identity of behaviour and appearance of male and female leads to mistakes; abortive courtship excursions are often made by two fish of the same sex. However, courtship is very rapid and such errors waste little time. They would be more serious if sexual products were shed and, in view of the synchrony of the spawning, it is not immediately clear what prevents this. Perhaps the behaviour of the two sexes is after all not quite identical; for instance, the male pushes vigorously against the female's side, sometimes tilting her over.

This is a very simple type of courtship, with few complications. There is no aggression to be overcome and no nest to be displayed. However, whitefish do have to spawn over a certain type of bottom and, in the absence of a territory, fish must indicate both that they have found a suitable locality and also that they are ready to spawn there. This is certainly the function of the sailing display. The female turbot *Scophthalmus maximus* another non-territorial fish whose eggs lie on the bottom, gives a similar display; the cod *Gadus morrhua*, whose eggs float on the surface, has no such display.

The courtship of whitefish is typical of a simultaneous spawner; the mutual trembling, building up to a climax, is the significant feature. This type of mating requires much closer timing than any other, and identical mutual displays seem essential. The involvement of both partners in the climactic display is much greater than in, for example, the jewel fish. This too has mutual displays, but the female spawns first and the male can be stimulated to fertilize by the sight of the eggs. In fact, sometimes female jewel fish do spawn when the male is temporarily resting: he then comes quickly to fertilize the eggs.

The guppy

The guppy *Lebistes reticulatus* is a freshwater fish, native to Trinidad and the north-east coast of South America. It has internal fertilization and is viviparous; the young are independent as soon as they are born; there is no parental care and no territory. Guppies live in large schools and are social throughout their lives, even at mating time, in contrast to the other fish described. Their courtship has been studied by Baerends *et al.* [10].

The absence of territorial behaviour does not, in this case, mean the absence of hostility. Unlike most schooling fish, male guppies are very aggressive, particularly when in sexual mood. Indeed their pugnacity seems to minimize interference in courtship and mating, both of which occur within the group. Inexperienced males chase and attempt to court fish of either sex but, since males turn and attack them, they quickly learn to limit this behaviour to females. To aid this learning there is considerable sexual dimorphism; males are brightly coloured all the year round; females are very uniform in colour. In fact, female guppies closely resemble the females of other species which live in the same waters, and this is another source of error. Usually, the fact that the fish congregate in schools of their own species prevents males from courting females of the wrong species. Guppy males do court females of the species *Micropoecilia parae* and *Poecilia vivipara* if placed in the same aquarium; indeed they seem at first to prefer *Micropoecilia* females to their own. However, they quickly learn to concentrate on their own females, probably because they alone respond correctly.

Male guppies vary a great deal in colour. These variations have been much studied by geneticists and have been made more complex by fish fanciers, but so far little has been discovered of their effects upon other fish. However, at breeding time, several series of black markings appear on the bodies of the males, and these to a large extent obscure the colour differences; hence the latter may be of no significance in breeding. The black markings are similar in all males but come and go with changing mood. (The word 'mood' is used here, and throughout this book, to imply a high tendency to perform a particular behaviour pattern. Hence males are said to be in sexual mood when they court a lot and have a particular group of markings; but in aggressive mood when they fight a lot and are differently marked.)

Guppy courtship has two main stages, corresponding to two

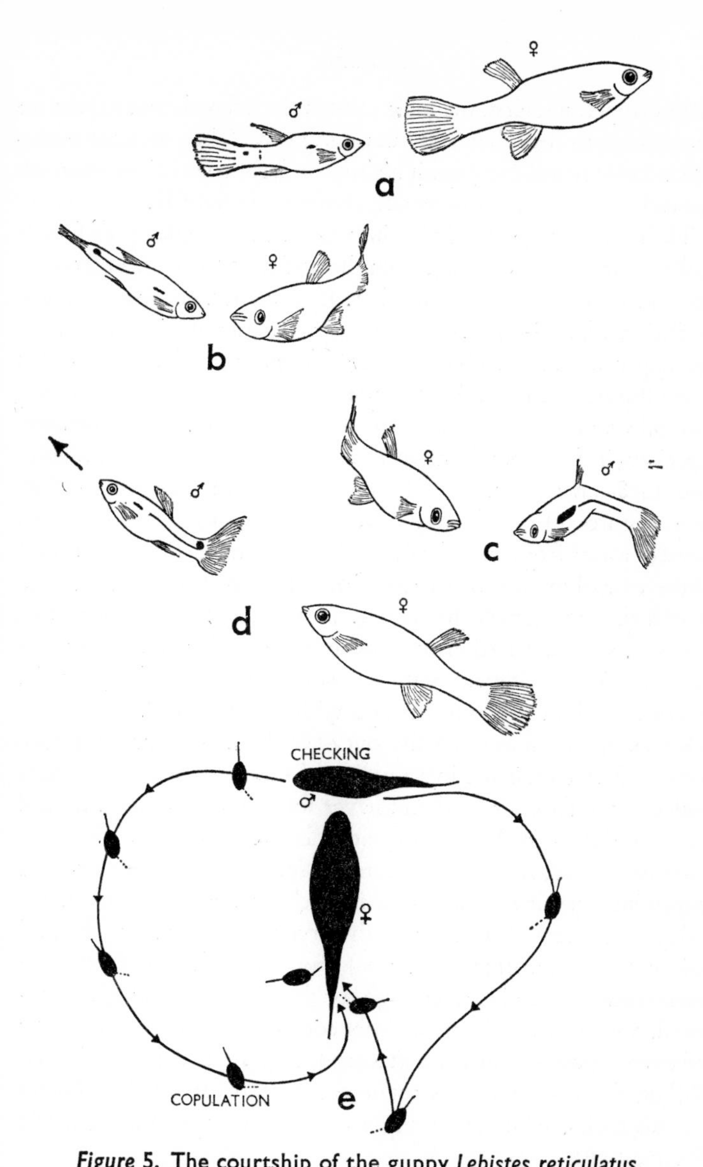

Figure 5. The courtship of the guppy *Lebistes reticulatus*. (After Baerends, Brouwer and Waterbolk [10].)
(*a*) Following; (*b*) luring; (*c*) S-curve; (*d*) jumping; (*e*) checking and copulation.

distinct functions. First, the male must entice a female away from the group so as to court her without interference. Then he must stimulate her to stay still and permit his approach for copulation. Baerends has called these stages *leading* and *checking* respectively.

The leading stage usually begins when a male singles out a female and follows her (figure 5*a*). Soon he attempts to swim in front of her and to turn so that he faces her. If successful, he postures in this position with folded fins, and then begins a series of alternate retreats and approaches, still facing her and swimming jerkily. This is called *luring* (figure 5*b*). It is as if he were trying to draw her towards him; and indeed his behaviour sometimes has this effect. Sooner or later, especially if she responds, he turns right round and throws his body into an S-*curve* (figure 5*c*). From this position comes a startling *jump* (figure 5*d*) which may cover a distance of 10 cm or more, directly away from the female. This is the most conspicuous part of the courtship, and is perhaps more effective than anything else in inducing the female to follow. Jumps and S-curves are often repeated, so that a responsive female is led well away. Sometimes successive jumps alternate with a posturing in front of the female, this time at right angles to her and with all fins spread.

Checking develops gradually out of leading. A leading sequence occurs which culminates in a modified S-display orientated at right angles to the female's head and never followed by a jump. Instead the male shimmers slowly in front of her head, moving forwards. Then he returns, sometimes reversing, sometimes turning first and swimming forwards again. This may check the female; if so, the male sweeps round her in a smooth curve to approach from behind and attempt copulation (figure 5*e*). For successful copulation, the female must stay still long enough for the gonopodium to be introduced. Sometimes males make brief contact with the female's body and even deposit sperm without introducing it into her genital tract. Willing females sometimes rotate the ventral part of their bodies towards males, while unwilling ones turn away with a slight but effective twist.

The complexities of this courtship are not easy to explain. To some extent the sequence of movements depends on the female's responses; her failure to follow or stay still at the appropriate moment may cause a male to repeat a phase of the courtship. There seems to be a relationship also with changing internal states in the male, as indicated both by his behaviour and by changes in the patterns of black markings referred to earlier. In particular, males tend to be aggressive towards their females (and show the markings associated with aggression) during the early phases of courtship. In later phases males rarely attack (they more often flee) and markings associated with sexual mood tend to predominate. Further evidence on display and mood in this fish will be given in a later chapter.

Explanations of this complex courtship, whether in terms of function or of origin, seem to be related to the guppy's social habits. Because of crowding, it is necessary for a male to entice a female away for mating; because of the hostility within the group, he seems to be not only sexually inclined towards his female but also 'pugnacious' and 'afraid'. The swordtail *Xiphophorus helleri* is also a social fish which resembles the guppy in having internal fertilization and in lacking a territory. Yet its courtship, which has been described by Morris [193], is remarkably simple. It is interesting to compare these two.

The whole of the swordtail's courtship bears a remarkable resemblance to the checking phase of the guppy's. A male follows a female, at first keeping his distance but finally darting suddenly ahead displaying his tail and flank. Then he shimmers slowly backwards for a copulation attempt. Failing this he sweeps again across her escape path to repeat his display and so on. Reasons for the greater simplicity of swordtail displays are suggested by differences in the social system. Male swordtails do not fight continuously; they tend to keep their distance after initial skirmishes. Baerends believes this represents incipient territorialism [8]. Whatever its nature it has something of the functions of a territory in reducing interference. Hence male swordtails do not 'need' a luring phase in their courtship; they can court females within the group. At the same time, the aggressive reactions between male swordtails appear to be stimulated mainly by the sight of their distinctive tails: females lacking these are rarely attacked. Hence some of the complicating factors present in the guppy (especially during luring) are absent in the swordtail.

Survey

These descriptions suggest some typical features of courtship and the factors which influence them.

Where there is territorial behaviour, courtship has a typical opening phase. A female seeks out a male on his territory attracted by his bright colours and conspicuous behaviour (which also warn off other males). She is challenged (or attacked) and responds with a signal indicating her sex and willingness to mate. Occasionally she fights back, but usually she signals too. Signals may resemble appeasement postures adopted by defeated rivals or, occasionally, they are activities performed exclusively by the female before mating. For example, the female char *Salmo alpinus* digs a nest in an exaggerated fashion within the male's territory [84]. In addition, females often lack the bright nuptial colours of the males. Both signals and appearance tend to check the male's hostility so that it is gradually replaced by sexual behaviour; nevertheless attacks are common in early courtship.

Hostility, bright nuptial colours, and female signals all occur also in non-territorial fish, although not always in the same context. There is fighting between the males of many social species and most of them have bright nuptial colours which signal their hostility. For other species similar bright colours seem to be important mainly in attracting females. This is particularly so where fish spawn communally in special places. For example, a school of male stone-rollers *Campostoma anomalum* (North American freshwater Cyprinids), clears an area of debris and remains there, in a close group, until females are attracted by the mass of colour (the males have orange and black dorsal fins) [200, 251]. Finally, non-territorial females may give special signals in courtship, like territorial ones. However, these are rarely 'appeasing'. Usually they indicate willingness to mate. Goldfish *Carassius auratus* females make vibratory movements when ready to spawn and these induce following in males [32, 200]. Zebra-fish *Brachydanio rerio* females, in a similar mood, begin to drive their males; this stimulates the latter to turn and drive them instead [6, 200].

Differences in *mating habits* tend to affect the closing rather than the opening stages of courtship. The most obvious type is that of the simultaneous spawner, of which the final mating displays consist of mutual undulatory or trembling movements, accelerating to a climax when spawn and semen are shed. By contrast, where there is internal fertilization, mating displays are usually 'asymmetrical': the male displays actively and the female is passive; he needs only to induce her to stay still and permit his approach, for she does not spawn at this stage. As a further point of contrast, tactile stimuli are rarely used, for females appear to avoid contact until they are completely willing.

Fish with external fertilization, which are not simultaneous spawners, have more diverse displays. Usually the female must be induced to lay her eggs in a certain place when the male is at hand to fertilize them. Hence one would expect the main function of the male's display to be the stimulation of spawning at that time and place. Sometimes only the male displays, thus exciting the female, for the sight of the eggs may be sufficient to induce spawning in him. However, where stimulation is associated with mutual spawning site displays, as in the jewel fish, then a 'symmetrical' courtship may occur, in which male and female behave alike. The stimuli presented by fish of this type also vary. Sometimes they are contact stimuli, as in the 'trembling' of the male stickleback; sometimes they are visual, as in the sea-horse *Hippocampus* which postures in front of his female to invite her to deposit eggs in his pouch [281, 200]. Zebra-fish appear to chase females until they lay.

Parental behaviour particularly affects courtship where there is special behaviour at mating, relevant to the welfare of the young. When parents stay together, after mating, to care for the young, a comparatively lengthy courtship is the rule, wherein the fish become acquainted. In addition, nest or spawning-site displays, or luring behaviour, are common features of early courtship. Sometimes fish posture over suitable areas, like the 'sailing' whitefish, or collect there as in stone-rollers. Nest-building, when it occurs, usually precedes mating (for the eggs must be laid in the nest) and may be performed by either sex.

2: Birds

Courtship is perhaps more complex and more diverse among birds than in any other group. Yet they are rather uniform in their reproductive habits. The large majority build nests within defended regions and show considerable parental care. All have internal insemination and all lay eggs. Obviously diversity in courtship cannot be attributed to the presence or absence of these features.

However, bird behaviour, in general, is very diverse. First, territorial behaviour itself varies greatly. Territories may be large or small, held for a short period, or held permanently. On the one hand, colonial birds tend to have closely grouped territories, with much interaction between neighbours. On the other, some birds have territories so large that they rarely see other pairs. By contrast, when territories are permanent, neighbours may know one another and their respective territories very well, and boundary disputes may be few. Hence the degree of hostility associated with territorial behaviour covers a wide range. In addition, some birds pair, not in their territories but in the flock before the territory is established. The typical 'hostile' opening phase of courtship (or pairing) may then be lacking. Other forms of territory include the special pairing territories of some gulls (Laridae) and the special mating territories of the ruffs *Philomachus*, blackcocks *Lyrurus* and other polygamous birds. Some polygamous birds hold no territories at all. All these variations have their repercussions on courtship [115].

Secondly, social relationships are better developed and more diverse in birds than in fish. Almost all birds are social at some period of their lives, and even those which simply flock in winter show much more complex interactions within the group than do most schooling fish. Flock members communicate with calls and other signals, and these often appear in courtship too. For it is normal for mated pairs to recognize each other as individuals (as would be expected from their shared parental behaviour) and they tend to react to one another as flock partners. Permanently social birds, such as jackdaws *Corvus monedula*, develop communication systems even further and many have 'hierarchical' societies [158, 161, 164]. Social birds, too, often develop communal displays, for it seems that the excitement of many birds displaying together is very stimulating to females.

Thirdly, habitat preferences influence the courtship of birds, as of fish, but again there is more diversity. Among fish, static displays can occur only in relatively still waters, and all visual displays must occur close to the mate because it is difficult for fish to see for long distances in water. 'Close-up' displays also occur in birds living in thick woodlands, but open country and sea shores lend themselves to spectacular flight displays. Water birds have other possibilities. Sometimes the display may be affected by the length of the breeding season. Crook [70] has suggested that display differences between two species of weaver birds *Ploceus* relate to the fact that one inhabits arid regions (with short seasons), the other more humid areas (with longer ones). Other variations occur because of the many different types of movements possessed by different species of birds. Some fly more than others, some tend to run along the ground, some flick their tails or toss their heads when about to take off, some preen or bathe in a conspicuous way. These different habits may be reflected in courtship.

Finally, besides these sources of diversity, there is a feature of bird behaviour, entirely lacking in fish, which confers certain characteristics upon bird courtship, that is, the possession of song. This is an additional and highly effective form of advertisement; it attracts females and signals the ownership of a territory. Song can operate over long distances and can be heard even when the singer is hidden; hence birds do not normally need conspicuous visual displays at pairing time (although these are occasionally found). Song is also highly specific. Young males usually inherit a tendency to sing a certain type of song, which is later perfected by learning from parents and neighbours (the degree of learning varies with the species [265]).

Hence song remains uniform within the species, and females with a tendency to respond to that song will rarely stray into territories belonging to birds of the wrong species. Song thus prevents cross-mating, and is often all that is needed for this.

The yellow-hammer

The yellow-hammer *Emberiza citronella* is a common resident British passerine and its courtship is typical of that of many song-birds. This description is taken from that of Andrew [2]. The birds winter in relatively peaceable flocks, and the males begin to separate to set up territories in February. Within their chosen areas they soon begin to sing from prominent stations, often hedgerows. Females then begin to visit them, perhaps attracted by their song and their gleaming yellow breasts.

Pair formation follows in the typical territorial fashion that has already been described for fish. Yellow-hammer females are not themselves territorial; but they are capable of aggressive behaviour towards other yellow-hammers, for this sometimes occurs in the flock. Visiting females therefore threaten males, fly up to them and supplant them, and sometimes fight them. This very bold behaviour, together with dull colouring and lack of song, seem sufficient to signal the visitor's sex since, although the male threatens and sometimes attacks, he is already treating the female differently from a male when he allows her to supplant him. Perhaps because of this, the female has no special 'appeasement' posture.

Females often visit several males, coming and going between neighbouring territories. Eventually they stay more and more with one male, and then his aggression quickly fades. Now he will join her on the ground, feeding and sometimes picking up nest material with her. Feeding or pecking together are common flock activities: it is as if each partner has adopted the other as a substitute for the rest of the flock, for during the following weeks they keep close together, feed together, leave and return together and give the tail flicks, flight calls and social calls which are typical signals within the flock.

The carrying of nest material which occurs at this time is often accompanied by signs of sexual activity in some of the postures of the male. The female is evidently recognized, not only as a flock substitute, but also as a potential mate, and at first this excites activities associated with mating. However, both sexual and nesting behaviour soon disappear, although no nest has been built and no mating has occurred. The two birds now live quietly together for a few weeks.

A prominent feature of the 'engagement' weeks are mock attacks made by the male upon the female, called sexual chases. The female is chased closely through the territory for a considerable time, until indeed the male often seems exhausted. The female does not seem to be very frightened and indeed is often the less agitated of the two, even though the male occasionally attacks her. Such chases are common in other birds too; they may help to stimulate the growth of the female's gonads.

True nesting and mating behaviour begin together early in May. They are closely linked. The male does not build, but the sight of the female stimulates him to pick up nest material and run with it to a nest-site which he has previously chosen. This is always well under cover, and hence needs to be pointed out in this way. Once there, the male displays with horizontal body, lowered bill and one or both wings extended and vibrating. There is also now a special call. These displays alternate with true sexual displays orientated about the female, not about the nest.

The male has two sexual displays. The first is the *fluffed run*. The male makes repeated short runs directly away from the female. He appears frightened and has fluffed feathers and a drooping tail and wings. Usually this activity ends in the collection of nest material, and indeed many features of the male's posture during the display suggest that he is about to do this. In particular he tends to jerk his head occasionally towards the ground. Feather fluffing, on the other hand, is most commonly seen in a frightened bird which cannot escape.

The fluffed run often alternates with the second display, the *bill-raised run*. Here the male again makes a series of runs, but this time towards the female. His posture, with erect body, raised bill and wings is that always adopted just before a male jumps on to a female's back, and if the female solicits he may actually do this. By contrast the fluffed run never ends in mounting.

The female's *soliciting* posture is also that normally adopted at coitus with additional wing quivering and sometimes a special call. A soliciting female always permits her male to mount, and never attacks him. In many ways the posture resembles that adopted by young birds when begging for food; perhaps this helps to make the male less afraid. However, although males rarely attempt to mount females unless they are soliciting, they do not invariably respond to such a signal of willingness. Sometimes, during a bill-raised run, a male may come right behind a soliciting female, and stand erect, but finally turn away.

This courtship is surprisingly elaborate and variable. The birds have been living together peaceably for some weeks; one would expect a simple, stimulating display on the part of the male, inducing the female to permit his approach. However, it seems difficult for the pair to come together. The female often pecks at the male as he approaches, and he is very hesitant, and very inclined to flee. In fact the relationship between the pair appears to change at this time, and this is apparent, too, in their other behaviour. The female may supplant the male at a feeding post or a favourite perch, and he retreats, although previously he would have been the aggressor.

Similar reactions between pairs at mating time are common among passerine birds. It has been suggested that they stem from a general fear of bodily contact, for many animals appear to avoid approaching too closely to one another on all occasions [110, 137, 270, 279]. However, although this is true of many birds, others, like the social spice finches *Lonchura*, actively seek bodily contact outside the breeding season [198]. Yet they too show avoidance when they are about to mate. More probably the female's aggression is always an expression of her tendency to fend off the male (if necessary by force) until she is quite ready to mate; the male's hesitancy is probably a reaction to this.

Whatever their causes, these difficulties tend to diminish after repeated coitus. The male becomes less hesitant and the female less aggressive. Occasionally coitus occurs without display. A female sometimes solicits spontaneously, particularly if her mate appears while she is carrying nest material, and, in response, he may fly straight to her and mount her.

It is perhaps of interest to compare this courtship with that of another territorial passerine which differs in being a social bird, nesting communally. This is the house-sparrow *Passer domesticus* L. Summers-Smith [261, 262, 263, 264] describes how breeding occurs in small compact colonies in gutters, spouts or eaves of houses and the territories comprise little more than the immediate vicinity of the nest. An adult bird is usually faithful, not only to its colony, but also to its mate and nesting site; indeed some roost at the nest-site all through winter. Hence colony members are well known to one another and often perform activities like feeding and dust-bathing together. This synchrony extends to courtship behaviour too, even to activities like nest-building and copulating, which are strictly between pairs. On the other hand, there is also some interference, for a courting pair sometimes attracts other males.

There are many resemblances between the courtships of the yellow-hammer and the sparrow. Pairing in the sparrow may occur only once in a lifetime, but otherwise the process is similar. The male of both species is aggressive to the female at pairing but becomes timid at mating. Both types of female are aggressive when courted, but solicit when they are ready; both types of male may respond to a soliciting female, sometimes without preliminary display, and the female does not attack once she has solicited.

The main differences appear during the 'engagement period'. Like the yellow-hammer, the sparrow rarely mates before late April or May; nevertheless it begins both nest-building and courting much earlier. Perhaps both these activities help to keep the pair together. Both require the association of the mates, for nest-building is co-operative in this species. This may be important, for there are many opportunities for unfaithfulness in so close a community; pairs, particularly new-formed ones, do, in fact, sometimes break up. The most striking difference between sparrow and yellow-hammer is undoubtedly in the communal displays of sparrows [261]. These usually begin with a chase between a single pair, male chasing female as in the sexual chase of the yellow-hammer. Other males, however, join in and collect around the female as soon as she lands, displaying to her exactly as in a solitary performance between mates, although the males constantly change positions. The females are often aggressive and the males usually disperse after about a minute. However, during such a display males tolerate one another's presence, whereas they drive intruders away when they are performing alone.

Hence many of the differences can be attributed to the greater interplay between sparrows. In particular, communal courting is evidently a social development of the sexual chase: it too may help to bring females into breeding condition. And the synchrony of courtship leads to considerable synchrony of breeding within a colony: the eggs are laid in all nests at about the same time.

The green heron

The green heron *Butorides virescens*, described by Meyerriecks [189], is of interest because, being solitary, it lacks the social interplay and flock behaviour found in the two birds already described. Its courtship also affords an example of spectacular aerial display.

The green heron is a member of the order containing storks, ibises, spoonbills, herons and flamingos (Ciconiiformes). All breed in close colonies, perhaps for protection, but some, including the green heron, are otherwise solitary. The green heron is also migratory, wintering in South America, Mexico and Central America and breeding near fresh and salt water in the eastern part of North America. Like most migratory birds, it sets up its territory quickly, for the birds must begin breeding soon after their arrival. Groups of males settle in a suitable spot (often an old breeding ground) and, after prospecting for a day or two, each adopts a rather large territory, usually with many tall trees for song posts and preferably an old nest in one of them.

The advertising call is a low-pitched single 'skow' given from a tree top with the bill pointing obliquely upwards. Females who are attracted do not immediately enter the territory but perch outside, giving a slightly different call, or 'skeow'. Soon 'skow-skeow' rallies develop, followed by a typical intrusion into the territory by the female. She is greeted by attack, but is unusually bold and persistent: she stays her ground although she does not, herself, attack. Gradually the male's hostility declines and he accepts her as a partner.

The engagement period is shorter than that of the yellow-hammer or the house-sparrow and it is occupied, not with flock behaviour like that of the former bird, nor with nest-building and courting as in the sparrow, but with complex aerial nest-site displays. The nest-site displays of most birds are brief and their elaboration in this case seems to be related to the reluctance of male herons to allow any birds, even their potential mates, to approach their nests. This, in turn, appears to be related to the unusual fact that eventually the male's territory shrinks to include only the nest and its immediate vicinity. While the male yields other ground rather willingly, he retains a strong tendency to defend this particular spot even from females.

There are three main aerial displays. The first is the *circle flight* in

Figure 6. The flap flight display of the male green heron *Butorides virescens*. This is the most extreme of the nest-site displays. (From Meyerriecks [189].)

which one of the birds, male or female, flies in a circle of 20–50 metres diameter, landing back where it started. The flight is quite normal, but the bird calls 'skow' or 'skeow' as it flies. Birds may alternate in this display or even perform it together, but it soon changes to the second, or *crook-neck*, display. This resembles the first, apart from differences of posture and flight. It looks as if the bird is only half-hearted about flying: it does not fully adopt the flight posture, its neck is not fully retracted but kinked, its crest is raised and its legs are dangled as if in preparation for landing. In addition, its flight is abnormally undulating, with slow deep flaps of the wings. The third display is called the *flap flight* (figure 6) and is performed mainly by males. The flight is now never circular, but is directed at the female from the male's perch (usually the nest tree). The hesitant flying is even further elaborated. The wings beat so deeply that they seem to touch below the body, the neck is even more crooked, the legs dangle, the scapular plumes are raised and sometimes the eyes bulge. The bird lurches clumsily through the air and lands near the female.

The last two displays exhibit the main colour features of the birds. The neck is chestnut, and the dark green scapular plumes form a striking background to it; the legs are orange-red, and the eyes yellow, sometimes changing to orange during the flap flight display. Male and female are similarly coloured throughout the year, but the male is brighter and more lustrous in the breeding season.

These three displays are best understood as inhibited excursions away from the nest on the part of the male, inhibited excursions towards it by the female. At first the male flies mainly in a circle from his nest tree and back, alternating with attacks upon the female. His circles sometimes start towards the female, but he seems quickly drawn back to his nest as if he cannot bear to leave it. The female's flights usually begin towards the nest tree but veer away. From time to time she does try to land there, but is driven off.

During the displays the male progressively changes his behaviour from hostility towards the female to more tolerant and even sexual behaviour. He attacks her less often and more half-heartedly during the two later displays, and the female seems to recognize the diminishing hostility. For, when a supplanting attack is made during a crook-

(*a*)

Figure 7. (a) The stretch display of the male green heron. This is a sexual display which can be contrasted with (*b*) the threat display of the same bird.
(From Meyerriecks [189].)

neck display, she usually shifts her perch only a few metres and immediately returns. The flap flight display itself may be considered as an inhibited supplanting attack, yet she rarely moves away at all in response and may even be stimulated to approach the nest afterwards.

The so-called sexual displays, which entice the female to the nest, now begin to alternate with the aerial displays. These are static and

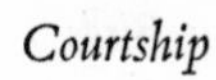

Figure 7. (b)

cannot appear until the male permits the female's close approach. The first is the *stretch* display (figure 7*a*). Head and neck are both stretched upwards, the fan of the interscapular plumes is displayed and the bird sways from side to side, uttering a soft aaroo-aaroo call. Many features of this posture seem to symbolize the negation of hostility. In threat the head and neck point forwards and the cry is harsh (figure 7*b*), whereas in stretch the main weapon, the beak, is pointed up and the call is soft. The female responds by approaching the nest more boldly and, after repeated stretches, she finally enters it.

The second sexual display alternates with stretch. It is called the *snap* (figure 8), for the bill is snapped with a sharp click while the head is stretched forwards and downwards. These movements resemble those used for snapping up fish for food or twigs for nest-building. Bobbing and bowing movements may be added in intense displays, and these too look like nest-building postures.

The culmination of these displays is the entry of the female into the nest (or nest-site). Both birds seem very tense, and the male may threaten the female if she approaches quickly; yet he may flee if she makes a sudden movement after she has entered. There may be many attempts before nest entry is finally achieved, but thereafter the male's behaviour changes abruptly. There are no more aerial displays, no more snap displays, no more attacks. Coitus and nest-building follow almost at once.

Before the first coitus there is a soliciting display by the female, identical to the stretch display of the male. The male sometimes

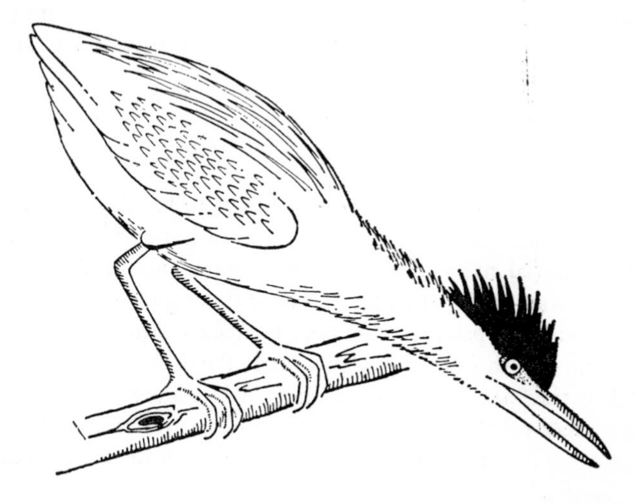

Figure 8. The snap display of the green heron: another sexual display. (From Meyerriecks [189].)

stretches too, but more often there is mutual feather nibbling, billing and bill snapping (close to the mate's head). Both birds show signs of timidity just before mating and the situation is probably not very different from that in the passerines, although the female never attacks the male. In later copulations these preliminaries drop out, so that there is no pre-coitional activity; the female solicits and the male mounts.

This complex courtship can, at least in part, be understood in terms of the unusual circumstances in which it occurs. The large territory, which gradually shrinks, arises from both migratory and solitary habits. The first arrivals need many song posts for advertisement to help other birds to find them. As the colony grows this need diminishes. The consequent gradual yielding of other ground probably promotes an unusually vigorous defence of the nest-site itself. As suggested, the aerial nest-site displays are perhaps related to this. The unusual mutual 'identification' ceremony, which occurs before the female enters the territory, may also be attributed to solitary habits. The birds do not know one another. In addition, no group behaviour or group signals have developed in these birds. There is none to be found in courtship.

The mallard

The mallard *Anas platyrhynchos* belongs to the family Anatidae, which includes also the geese and swans. It is a good example of a non-territorial bird which courts on the water and spends most of its life in a community. Pairing and even the engagement period are communal, and pairs separate only for coitus. After that the ducks nest, and rear their young, alone, while the drakes stay together in small parties nearby. These breeding habits are exceptional; even some other members of the Anatidae hold conventional territories. Lorenz has described and compared the displays of many species in this family [162, 163].

The male mallard wears his nuptial plumage from September until June, and pairing occurs very early, usually in late October or November. Drakes are very different indeed from the mottled brown ducks: their most conspicuous features are the dark green head, white collar, purple breast and four black tail covers which curl upwards.

The communal pairing ceremonies are remarkable in being mainly male exhibitions: at first the presence or absence of a female hardly matters. Females occasionally stimulate the displays by an activity called *nod-swimming* (figure 9*a*), but they remain onlookers during most of the performance. This too is exceptional, since ducks of other species, closely related to the mallard, court more personally, although still communally.

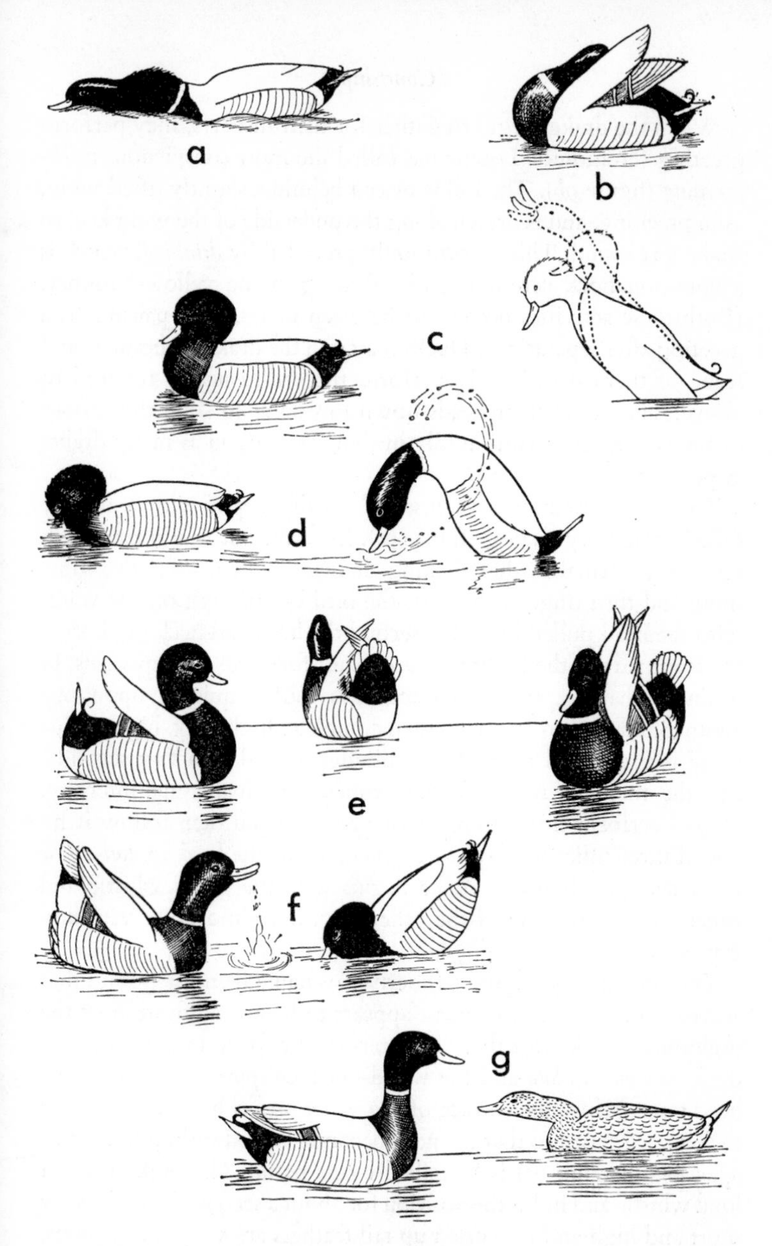

Figure 9. Courtship displays of the mallard *Anas platyrhynchos*. (After Lorenz [162].)

(*a*) Nod-swimming; (*b*) mock-preening; (*c*) preliminary shaking; (*d*) grunt whistle; (*e*) head-up, tail-up; (*f*) down-up; (*g*) 'pumping' of drake and duck before mating.

D

When the drakes begin to gather for their displays, they perform greeting ceremonies. Lorenz has called the most conspicuous *mock-preening* (figure 9*b*). The bill is placed behind a slightly lifted wing, as in preening, and is drawn along the underside of the wing keel to make a rr sound. This is commonly preceded by *drinking*, which is a common flock ceremony, like feeding in the yellow-hammer. (Both these activities occur also between mates, after pairing, as a greeting after separation.) Ducks may join the drake gatherings, and some of them occasionally perform the *nod-swimming* referred to above. They hold their heads down low so as to skim the surface of the water, and swim rapidly in short arcs round as many drakes as possible.

The male ceremonies themselves usually begin with a display called *preliminary shaking* (figure 9*c*). In form, but not in timing, it resembles normal feather shaking. It is in two parts – high swimming and thrusting. In the first, the bird swims high on the water with the head pulled in; in the second the head and neck (and often the front part of the body too) are thrust forwards and upwards. In ordinary shaking, these two activities follow quickly upon one another and the movement is then complete. In display, high swimming is prolonged (for several minutes instead of a few seconds) and the thrust is repeated three times with increasing intensity. Drakes perform this activity at one another and then follow it by one of three other displays – the *grunt whistle*, the *head-up, tail-up* or the *down-up*. Often one drake begins with the grunt whistle and others reply with one of the other two, but almost any variation can occur.

The *grunt whistle* (figure 9*d*) derives its name from the sound that accompanies it; the movement appears to be a modification of the preliminary shake (for there are intermediate forms in the mandarin duck *Aix galericulata* and the wood-duck *Lampronessa sponsa*). The head starts below the surface of the water and is held there during the upward thrust so that the neck becomes arched right over. *Head-up, tail-up* (figure 9*e*) is described by its name. The drake gives a loud whistle and holds the position for about a second; he looks very short and high and his curled up tail feathers are very conspicuous. On his subsiding, the head may point at a special duck. This is the only display of the courtship directed to a duck. After it the drake will nod-swim, often around the same duck, ending by displaying the back of his head to her. *Down-up* (figure 9*f*) looks very much like drinking. The bill is dipped and raised quickly and jerkily, often so as to send up a fountain of water. However, no drinking occurs; instead the drake whistles and afterwards gives a two-syllable quack-quack which is a familiar flock greeting note.

During these displays movements of the feathers bring about special colour effects. During preliminary shaking the head feathers are erected so that they look black instead of green. During head-up, tail-up, head feathers are sleek at the side but lifted at the back; during nod-swimming only a small patch of neck feathers is erected. This has special significance in the display of the back of the head to the female, for it shows as a bunch of black feathers framed by green ones. In related species there are permanent feather arrangements made conspicuous by fixed colour patterns.

Communal displays continue from September to January, during which time each drake becomes attached to a particular duck. It may be the ducks which choose the drakes, for they tend to nod-swim round special drakes between or even during the ceremonies. Hence when a male favours a particular duck during head-up, tail-up he may only be accepting a duck who has already chosen him. Paired birds normally continue to participate in the communal ceremonies.

After pairing time the flock lives peaceably together, usually in large groups on ponds or on the sea. Mates tend to stay near one another, however, and occasionally drakes may be seen driving their ducks and chasing other drakes away. This is usually a preliminary to coitus, which occurs only between pairs which have withdrawn a little from the others.

Coitus reaches a peak in early spring and is preceded by a mutual display (figure 9*g*), usually begun by the female. The head is moved rhythmically, up slowly and then down with a jerk (called pumping). The male's display is less intense. It is performed facing the female and always out of phase with her so that his head is up when hers is down. Eventually he swims round and treads her.

When the ducks go off to build nests (usually on islands), their drakes stand nearby in small parties. They now tend to chase any duck which approaches, and attempt to tread her. Such chases are communal, and the ducks tend to flee from them and do not return. This behaviour serves to keep other breeding females away and hence prevents overcrowding.

The main feature of this courtship is the communal pairing. It seems that many birds are greatly stimulated by communal displays and one would expect such displays to be elaborated in a non-territorial, gregarious bird with little aggression between flock members. There is, in fact, a predominance of flock behaviour; the whole of the pairing display seems to be an elaboration of the preening and drinking behaviour used as greeting ceremonies within the flock. 'Aggression' occurs only in connection with the isolationist desires of the pair at mating time; there is little hostility between the mates. The withdrawal of the pair for coitus reduces interference; the behaviour of males at breeding time achieves some spacing of nests and hence a better food supply and better concealment from predators for the young. The territories of territory-holding birds often meet such needs.

The mutual displays preceding coitus are peculiar but not unique. Mutual preening occurs, as we saw, in the heron; and gulls (Laridae) also have mutual pre-copulatory displays, in this case consisting of food-begging [273].

It is of interest to compare mallard courtship with that of two other highly social birds, the domestic fowl *Gallus gallus* and the blackcock *Lyrurus tetrix britannicus*, both belonging to the Order Galliformes (which includes all game birds). Both resemble ducks in that the females nest alone after mating; hence nesting behaviour does not coincide with mating and there are no nest-site displays. However, they differ from ducks in being polygamous: pairs do not stay together, for males mate many times in quick succession. In these two species, then, the association between the mates is very brief, pairing and mating occurring in quick succession.

The fowl, studied by Wood-Gush [291, 292], has an elaborate peck-order which requires all the birds to be constantly aggressive. Females take their place in the system and hence are liable for attack even though they are readily distinguishable from cocks. Like many other birds, the cock changes his attitude towards a hen when in mating mood; he allows her to pull at his feathers and even to peck him without reprisal. In other words, he becomes more timid towards her and she more aggressive towards him. Before coitus a cock will try to entice females by group activities such as feeding and dust-bathing. When performed by any member of the group these ordinarily stimulate others to join in. Nevertheless his courtship consists almost entirely of threatening and aggressive movements; indeed all the pre-coital displays occur in fighting situations too; and two of them, strutting and waltzing, certainly have an intimidating effect during fights. However, these acts stimulate the female sexually: Wood-Gush found that the cocks who chased and waltzed and strutted the most, obtained more soliciting crouches from hens than did more sluggish and less intimidating birds [292].

The courtship of cocks is, then, a mixture of enticement and aggression. The enticing group activities are reminiscent of duck pairing behaviour, but the aggressive nature of the cock's display is a point of difference evidently related to the nature of his society. Because these are mating as well as pairing displays, it is necessary for each cock to avoid interference and other cocks are fiercely driven away.

Blackcocks, by contrast, hold communal displays; these have been studied by Selous [247], Lack [140] and Koivisto [136]. The males gather together in special areas, called leks, and attract females by displaying their lyre-shaped tails and occasionally jumping in the air or making short flights. However, these gatherings are not amicable, like those of ducks, and there is no group behaviour; in fact each bird is defending a small territory and there is much threatening and bickering, even occasional fierce fights. Recent observations suggest that the strongest (and most aggressive) males hold the 'best' territories in the centre of the group and obtain most of the females [136]. When a female enters a male's territory he courts her with a distinct display which is not aggressive. He slowly circles her, making an occasional bow in front of her until she solicits. In this case, then, interference in mating is avoided by territorial behaviour although the close grouping of the territories allows the initial 'attraction' displays to be, in a sense, communal.

Survey

The examples given illustrate the diversity of bird courtship; but they are not representative, for too much emphasis has been given to the atypical. Very many birds pair, mate and breed on a single territory, like the yellow-hammer. The courtship of these species has much in common with that of fish, such as the jewel fish, which do likewise. Usually, however, pairs of birds are associated for much longer before mating than fish. There are three phases in courtship displays of this type.

During *pairing* prospective mates come together; thereafter they stay together at least until mating. In territorial species, the song of the male is both advertisement and threat. So, too, very often, are flight displays. Advertising song-flights occur in the skylark *Alauda arvensis*, curlew *Numenius arquata*, snow-bunting *Plectrophenax nivalis*, tree-pipit *Anthus trivialis* and many others.

Visiting females are often attacked, but, if the same female persists in her visits, this behaviour is soon replaced by more 'friendly' behaviour. Frequently there are 'group' activities, sometimes incomplete nesting or sexual behaviour and occasionally parental behaviour. Gull females of many species beg for food in the manner of chicks

Figure 10. Food-begging in the common gull *Larus canus*.
(a) Female begging, male watching: this male may feed her or join in (*b*) mutual food-begging: this is a prelude to mating.
(From Weidmann [282].)

(figure 10) and males feed them [282]; terns *Sterna* have elaborate flight displays in which food-begging and courtship feeding play a prominent part [11, 73]. Sometimes there are mutual appeasement signals. Black-headed gulls *Larus ridibundus* 'head-flag', each bird turning its face from its mate and thus concealing the chief weapon (the beak) and the 'aggressive' black hood [275] (figure 33, p. 109). Appeasement signals are, however, less often used by females to avert male attacks, as they are by fish. The females seem to be sufficiently distinguished by their boldness, their appearance and their lack of song. Where males possess bright nuptial colours, females usually lack them.

Non-territorial birds also pair (unless they are polygamous), and they often hold communal pairing ceremonies as ducks do. These provide considerable additional excitement. Some territorial birds also pair within the flock, before setting up their territories. Further, many species perform communal antics before the beginning of the breeding season. These may have significance for pairing and perhaps for the development of the gonads. Noisy ceremonial assemblies of magpies *Pica pica* and jays *Garrulus glandarius* occur in late winter, when much chasing through the branches occurs [103, 260, 289]. The screaming parties of swifts *Apus apus* are similar, but these continue into the mating season and sometimes aerial coitus occurs during them [289]. Occasionally there are stereotyped and elaborate displays. Parties of falcons *Falco peregrinus* fly to and fro on the same beat for hours at a time [289]; guillemots *Cepphus grylla* collect in parties on the water to display with drill-like precision [76]. More remarkable still are the elaborate dances of cranes *Grus grus*, including figures of eight, springing, throwing objects and freezing. These occur in groups outside the breeding season and are continued as solitary performances later, in the presence of the mate [289].

Engagement may be lengthy and is often peaceful. Males tend to dominate females, but there is little fighting. Displays may be relatively few and are usually a continuation of the more 'friendly' pairing displays. Early nesting and precocious (although usually incomplete) sexual activities may occur, particularly among social birds when they perhaps help to keep the pair together. Sexual chases are also common, and social birds may develop them into communal mating displays (as in the house-sparrow). Nest-site displays may also occur during this period if they have not done so during pairing.

Coitus commonly occurs at the nest, and is often closely associated with nest-building. In many species both activities occur simultaneously at the end of the engagement period. Often, at mating time, the previous relationship of the pair is reversed. The female becomes aggressive, apparently defending herself from the male's approaches, and he, perhaps in consequence, is timid. Soliciting signals are commonly given by the receptive female, and these seem to overcome the male's timidity: the coital posture is normally adopted (typically a crouch), but there are often placatory displays too, such as the stretch posture of herons. Food-begging is also common even among birds which do not use it at pairing. By resembling a juvenile, a bird seems to reduce both fear and aggressive tendencies in another. The wing quivering of the yellow-hammer and the house-sparrow are typical food-begging movements. The male sparrow may offer food just before coitus; hobbies *Falco subbuteo* and rooks *Corvus frugilegus*

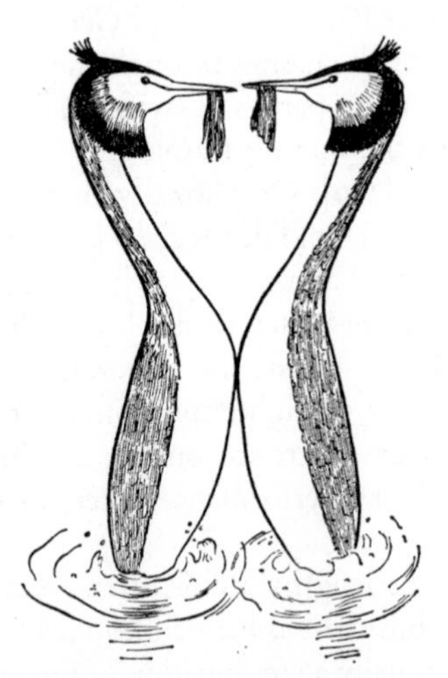

Figure 11. A mutual display of the great crested grebe *Podiceps cristatus*. The 'penguin dance' in which male and female rise up breast to breast and sway together after each diving for weeds. (After Huxley [123a], from Portman [215].)

perform ceremonial feeding at the nest, often followed by coition [289, 295].

Sometimes there are mutual displays, in which male copies female, as in ducks and gulls. Figure 11 is an example from the crested grebe *Podiceps cristatus*. More commonly, mutual feather-preening irregularly precedes mating. Birds such as pigeons, guillemots, ravens, rooks, cormorants and herons, provide examples. Sometimes, as in razor-bills *Alla torda* one of the birds adopts a so-called 'ecstatic' posture, with head stretched upwards and throat puffed while its mate preens [289].

The pre-coital displays of male birds vary considerably in duration and complexity. The more elaborate ones are usually found among polygamous birds like the peacock *Pavo cristatus*, where there has been no previous acquaintanceship between the pair; more stimulating displays presumably help males to mate more quickly and hence, perhaps, more often. Pre-coital displays of monogamous species may be limited to some of the brief mutual displays described above; occasionally they are omitted altogether. Examples have been given from the yellow-hammer and the green heron. Domestic cocks may also reduce or omit their displays when in a pen of hens with which they are familiar [292].

The problem of interference with mating seems to be an important one for most birds, for males seem to be attracted to couples in coitus. Even territorial birds, if they have closely grouped territories, may have to interrupt coital displays to chase away rivals. Lek birds (blackcocks and ruffs), in spite of their emphasis on communal mating displays, nevertheless hold territories for coitus; and many non-territorial birds, such as ducks and partridges *Perdix*, defend moving territories around their females at mating time.

To conclude, while some of the diversity and complexity of bird courtship, like that of fish, may be attributed to differences in their territorial, mating or breeding habits, a great deal is also an aspect of their complex social relationships. It has been seen how flock behaviour, flock displays, parental displays and appeasement displays feature in courtship. Other conspicuous acts typical of a species, genus or family may play a part. For instance, ducks have conspicuous preening behaviour and this features prominently in their courtship. More conspicuous still are the aerial displays: these are more commonly developed in courtship than the examples given suggest. Some birds, notably the raven *Corvus corax*, some of the harriers *Circus*, the falcon *Falco peregrinus* and the hobby *Falco subbuteo*, indulge in remarkable aerial acrobatics during courtship, tumblings diving, gambolling and somersaulting, often without pattern or sequence [289].

3: Arthropods

Among invertebrates courtship is common only among the more advanced arthropods and molluscs. Even in these groups it is not universal: it is possible to find closely related families, some with elaborate courtship, others lacking it entirely. Also one may find a complete range from very simple to very complex displays within one genus, such as that of *Drosophila*, the fruit-flies.

Most arthropods are non-territorial and the large majority are non-social. All those to be considered in this chapter have internal fertilization and are polygamous; in no case do the mates stay together for parental care. Hence breeding habits are rather uniform. On the other hand, arthropods are very diverse in their form and physiology and in their way of life. The differences of form between for instance, crabs and flies are obvious; differences in sense organs and sense physiology are just as great. Some groups are especially sensitive to chemical stimuli, others to tactile, vibrational, visual or auditory stimuli. Some groups entirely lack certain sense organs (like compound eyes or auditory receptors) which are very important in the lives of others. These factors account for some of the courtship differences among the arthropods.

Arthropod sense organs differ in many respects from those of vertebrates. The compound eyes have poor powers of definition or of distance vision but are particularly sensitive to movement. By contrast, organs of chemical sense (both taste and smell) are more acute and discriminatory than those of most vertebrates except mammals. In addition sense organs are not all concentrated on the head. These factors account for some of the differences between arthropod and vertebrate courtship.

In this chapter, considerable attention is paid to the experimental analysis of the signals involved in courtship. For most animals special stimuli (scents, colours or sounds), coming from potential mates, are important in initiating and often in directing courtship. They are called *releasing stimuli* and they may be limited so that animals, responding solely to them, are able to avoid courting unsuitable objects. The emphasis which has been put upon signals of this type in these descriptions of arthropod courtship does not indicate that they are of more importance in this group than in any other. Most vertebrates have highly specific sexual releasors too, and the discrimination of song by birds, for example, is more complex than the equivalent discrimination in grasshoppers and crickets. The emphasis stresses relative, rather than absolute, importance. In most vertebrates other features and functions of courtship are also present and often claim more attention; in arthropods recognition and guidance are dominant.

A cockroach

Nauphoeta cinerea is a cockroach of Asian origin but now widespread, particularly in America. It is chosen as an example of a very simple arthropod courtship with a rather unusual function. This has been described by Roth and Willis [235, 236].

Cockroaches have no problems regarding the meeting of mates: they have special responses to temperature, humidity and light which lead them to congregate in warm, damp, dark places. They also respond to species-characteristic odours; so that it is largely individuals of a single species which gather and stay together.

Sexually mature male *Nauphoeta* spend much time running around the group, touching antennae with the insects they encounter. They then move away unless they receive a chemical stimulus provided by the antennae of virgin females. This usually induces display and, in responding to it, males direct their courtship mainly to insects likely to accept them.

A stimulated male strokes his partner's antennae for a few seconds and is sometimes stroked in return. Then he turns his back on her, raises his wings through about ninety degrees, lowers his abdomen until the tip touches the ground, and poses thus for several seconds or occasionally as long as a minute. This posture exposes glands on the back of his abdomen, which are normally hidden by the wings, and which secrete a substance attractive to the female. She moves forwards, nibbling at the secretion, and hence climbs partly on to the male's back. Here she is in a convenient position for the male to stretch his abdomen, and grasp her genitalia with his claspers. If he is successful he twists his body from under hers and turns so that he is in line with her but facing in the opposite direction. This is the normal copulation position of cockroaches.

Few animals go to such lengths to manœuvre their females into a mating position, although many contrive to bring them to a standstill on the ground. It is not certain how valuable this device is to male cockroaches; perhaps they would otherwise find it difficult to push their way under females; perhaps occasionally they suffer from jostling in crowded colonies. It is possible, too, that either the secretion, or some other consequence of courtship, is sexually stimulating, for many females are initially unwilling but mate after several 'feedings' from the backs of males.

A fruit-fly

Drosophila is a genus of small fruit-flies feeding mainly on the yeasts found on decaying fruits and fungi. It has been much used in genetical and evolutionary studies and recently its behaviour has been studied too. There are many species, and most of them court, but the displays vary considerably, some being very simple, others remarkably elaborate.

Fruit-flies congregate at food and, although there are specific differences in food preferences, several species may be seen mixing freely at most good food sources. They stay near their food for long periods and the females lay their eggs there.

The courtship of *Drosophila melanogaster* has been described by Bastock and Manning [25]. A male always precedes courtship by tapping. He touches another insect anywhere on its body with his fore-legs. By this means he 'tastes' the insect, for he has taste receptors on the tarsi of these legs. If he perceives a 'wrong' taste (from certain other species) he may turn away without courting, but discrimination by this means is not very precise; he may still court members of several species, male or female.

There are three main components of the courtship. The first is *orientation* (figure 12*a*): the male stands close to the female, usually behind her, but sometimes facing any part of her body; he may circle round her but he always keeps close and facing, following her if she moves away. The second component, *vibration* (figure 12*b*), is performed while orientating. It consists of a vertical vibration of one

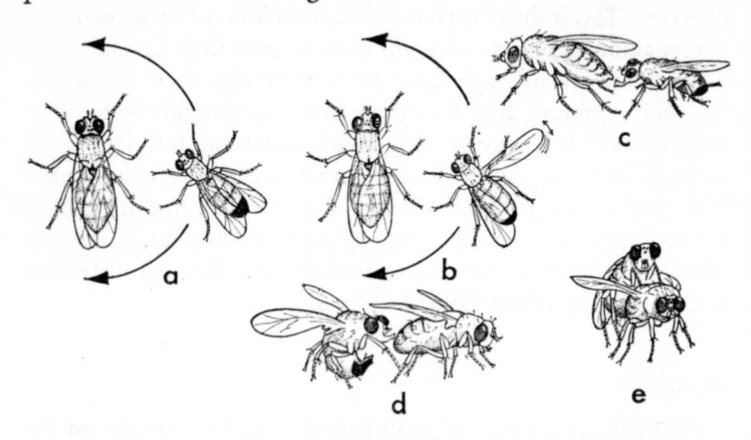

Figure 12. Courtship movements of the fruit-fly *Drosophila melanogaster*.
(After Manning [174].)
(*a*) Orientation; (*b*) vibration; (*c*) licking; (*d*) attempted copulation; (*e*) copulating pair.

wing which is held at right angles to the body. The wing is always that nearest the female's head and, if he circles her, he changes wings as he passes her head or tail. *Licking* (figure 12*c*), the third component, is performed during orientation and vibration. The male comes behind the female and licks her genitalia with a quick movement of his probocis. Usually licking is followed by an attempted copulation: the male mounts a female and curls his abdomen downwards and forwards to touch her genital region with his (figure 12*d*). A female may, however, resist such an attempt by escaping, by kicking, or by twisting her abdomen away. A willing female co-operates by spreading her wings and her genitalia (figure 12*e*).

It is unusual for a male's first sequence of orientation, vibration and licking (which takes a few seconds) to succeed. Even a willing female repulses a male at first. The male then returns to vibration or even to orientation, subsequently moving forward again in the sequence. Often, too, orientation and vibration alternate many times before a lick is made. Courtship can continue in this way sometimes without a break for a long period. Usually, however, a well-fed virgin female accepts a male in less than three minutes.

Fruit-flies of both sexes have strongly developed activities which drive away unwanted suitors. Males, courted by other males, flick their wings violently, kick and even turn to hit out with their front legs. A male already courting a female behaves similarly to another male who approaches; this behaviour tends to cut down interference, although not always successfully: strings or clusters of males may be seen pursuing one female.

Fertilized females repel courting males even more strongly. They may be settled on food for egg-laying and they rarely run away; instead, besides kicking and flicking, they may extrude the ovipositor and point it at the male. This nearly always checks him immediately. Virgin females repel males too, during their initial period of coyness, flicking, kicking and even extruding, but this behaviour is much less violent and it rarely drives a suitor away. However, virgins react strongly against males of other species. Females discriminate more sharply than do males and they repel males even of closely related species, often after the first tap. Hence this behaviour is important in preventing cross-mating.

Something is known of the stimulating qualities of courtship in *D. melanogaster* [24]. These insects mate just as well in the dark as in the light, hence visual stimuli are not essential. However, the wing display (vibration) seems to be very important, for wingless males take about twenty minutes to rouse a female, compared with about three minutes for normal males. Females without antennae do not discriminate between winged and wingless males; they take about twenty minutes to be aroused by either. It follows that females perceive a stimulus from this display with their antennae – in fact it is those portions of the antennae which perceive vibrations or air currents which are significant [210].

This independence of visual stimuli is not found in all species of fruit-fly. *D. simulans* described by Manning [171] is very closely related to *D. melanogaster*, yet it mates much less readily in the dark than in the light. At the same time the male's wing display contains less vibration and more wing scissoring, an activity described by its

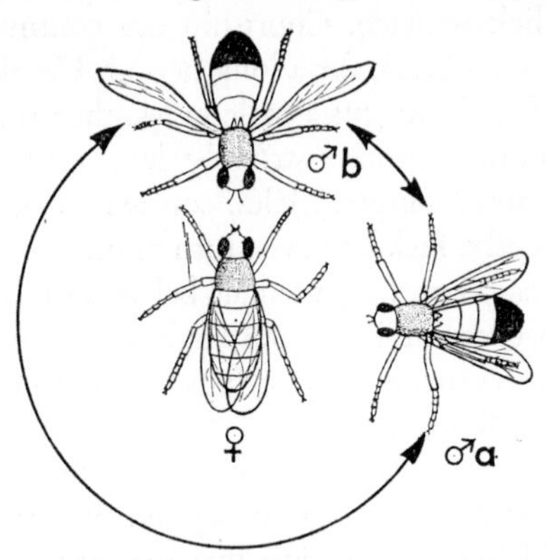

Figure 13. Scissoring in the fruit-fly *Drosophila simulans*. This movement commonly precedes vibration in this species but is rare in *Drosophila melanogaster*. Two positions of the male are shown. (*a*) The male may circle round the female, opening his wings only a little and 'scissoring' them rapidly; (*b*) when he reaches a point in front of the female he may adopt a static posture with his wings spread widely.
(After Manning [174].)

name and culminating in a static posture in front of the female (figure 13). This is probably visual in its effect. Maynard-Smith [186] and Brown [48] have described the more distantly related *D. subobscura* which does not mate at all in the dark. There is no wing vibration, and the bulk of the posturing takes place in front of the female, the male dancing from side to side in front of her head with his proboscis extended trying to touch hers, while she performs a reciprocal dance. Occasionally the male's wings are spread in a brief static display resembling that occurring at the end of scissoring in *D. simulans*.

These three species, *melanogaster*, *simulans* and *subobscura*, all belong to the subgenus *Sophophora*, the members of which respond to stimuli, visual or airborne, which can be detected from a distance. Spieth describes how a further difference is seen in species of the subgenus *Drosophila*, which respond mainly to contact stimuli in courtship – licking, leg-rubbing and leg-scratching. Wing displays are reduced in this group [253].

A very simple type of courtship indeed is that of *D. victoria*, a little known species of the subgenus *Pholadoris*. Here a male, after tapping, crouches behind a female with his abdomen curled under. From this position he suddenly attempts intromission. Mounting comes later, if intromission is successful. There is a very brief and slight extension of one wing together with an almost imperceptible vibration of it during the attempt at intromission, but not at any other time. There is no posturing or circling, and no licking [253].

These diverse courtships all have the function of inducing females to permit copulation. For, just as in birds and fish, even mature, virgin females appear initially 'coy'. In most species females give an acceptance posture when receptive, adopting the copulation position (wings raised, genitalia spread). However, males of some species try to mate without it, and the attempts are often successful.

The silver-washed fritillary

Butterflies are solitary insects; male and female must find and recognize one another before they can mate. The European silver-washed fritillary *Argynnis paphia* L. makes this problem easier by its restricted flying habits. This species, studied by Magnus [169], inhabits clearings and the edges of woods and tends to stay in one locality, rarely wandering far. The adults fly during July and August and, because of different temperature and humidity preferences, females fly mainly in the mornings, males in the afternoons. This would appear to reduce their chances of meeting; but there is a considerable period of overlap during the mid-morning, and a male who searches early in his period of activity is likely to find females who can easily be induced to settle.

Males search for females with a zig-zag flight which is quite distinct from a feeding flight. They also react to different objects on the two types of flight. When searching for flowers they react to yellow, green and blue objects; when looking for females, to yellow-orange objects, preferably fluttering ones. By experimenting with models, Magnus found that shape was unimportant in female searches, and size was influential only in the sense of the bigger the better (although models more than four times the size of a normal butterfly were not tested). Colour was very specific, and the fritillaries' own colour gave the best response, although models brighter than normal were even more attractive than a real butterfly. Fluttering was less specific (figure 14); the faster the better appeared to be the rule, up to speeds much faster than that of a normal

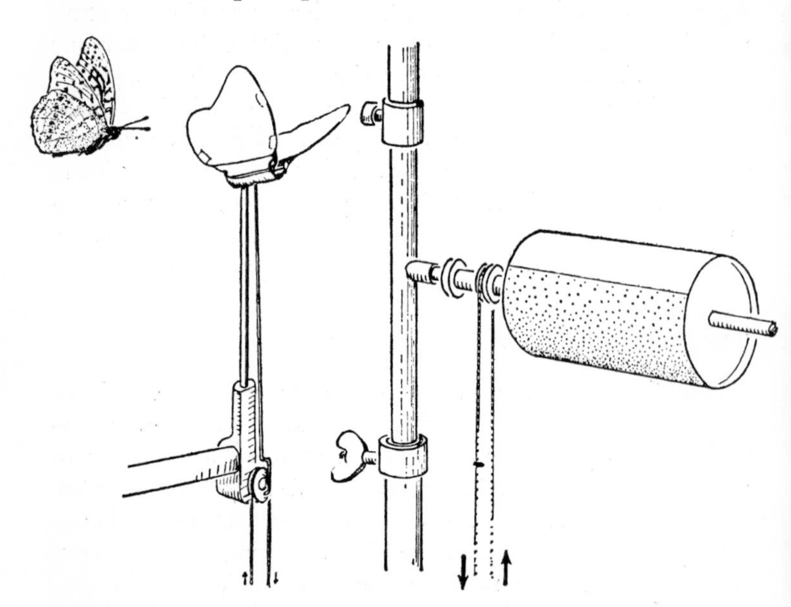

Figure 14. Fluttering dummy butterflies and rotating striped cylinders used by Magnus to test factors stimulating approach in male fritillaries *Argynnis paphia*. Both models were attractive and increased their attractiveness with speed of rotation up to the flicker fusion frequency of the fritillary's eye. Magnus concluded that the significant stimulus was the rapid alternation of colour and non-colour. (From Magnus [170].)

butterfly, until the flicker fusion frequency of the male's eye was reached [170].

Males make few mistakes in their sexual approaches. They do approach yellow fluttering leaves, but soon abandon them; and they do approach the tiny yellow Hesperid butterflies which fly in the same localities. However, they rarely follow these for long if the larger, and hence more attractive, fritillaries are around. The abundant Pierid butterflies are not the right colour.

After his approach, a male flies around the object which attracted him, in tight circles. Male fritillaries so approached respond with reciprocal circles; most other insects, including unripe and fertilized female fritillaries, fly away. Mature virgin fritillaries, however, flutter on the spot with a peculiar whirring movement. Males soon abandon all but the fluttering virgins and themselves take to

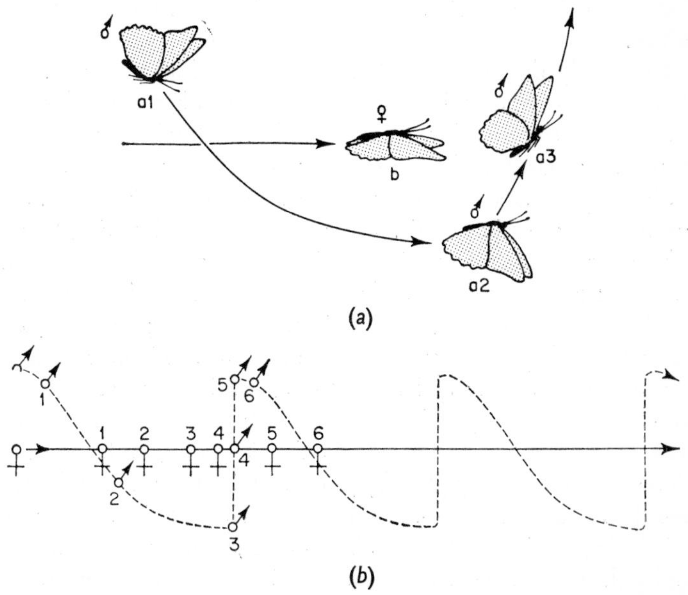

Figure 15. The pursuit flight of the male fritillary *Argynnis paphia*. A single swoop and upward dart by the male is pictured in *(a)* while three successive swoops and darts are indicated in (*b*). The female flies a straight course.
(After Magnus [169].)

fluttering, usually in a semi-circle at the same height. Thereupon the female flies away slowly on an exceptionally straight course, maintaining a constant height where possible. This course is quite unlike the usual undulating flight: it seems to act as a signal to the male, stimulating him to perform his remarkable pursuit flight (figure 15). As soon as he has caught up with her he glides down beneath her from a position slightly behind and above. Then, having overtaken her from below, he rises suddenly and almost vertically in front of her until he is above her head. She is then allowed to move ahead and the 'dance' is repeated. The male's regular upward darts must provide strong visual stimuli to the female's compound eyes; they also stimulate her mechanically and chemically, for he rises so close to her that she is often blown upwards. This gives a mechanical brake to her flight and at the same time permits her antennae to come into close contact with the scent glands on the male's fore-wings.

Eventually the female alights, usually on a leaf or flower. It is difficult to prove that the male's dance induces her to do this, but that is its presumed effect. Now ground courtship begins. This can be the only courtship if a male happens to encounter a female already settled. At first he flies in a semi-circle around her, as he did on first encountering her in the air, while she sits with fluttering wings, lifting her abdomen from time to time – probably to release scent. Soon he tries to gain a foothold where he can stand facing the side of her body; if he succeeds, the female stops her wing quivering and beats her half-open wings slowly up and down. This allows him to approach even closer and then he 'bows', clasping her body between his wings (figure 16). It is believed that by this means he funnels scent from his wing glands on to the female's body, for he suddenly closes his wings sharply (thereby probably releasing a cloud of scent), and then resumes his clasp. Meanwhile he vibrates his antennae and middle legs against the female's hind-wings, which are now held up and still. After the bow he continues this but transfers one antenna to the top of her head. Copulation is attempted by the male standing alongside the female, parallel to her and prodding her genital region with the tip of his abdomen. A receptive female stretches and turns her abdomen towards him.

Ground courtship is subject to interruption, caused mainly by the female flying away to feed, or to resettle nearby. Usually, however, a receptive female does not fly far and the male follows and resumes his courtship, although often starting again from the beginning.

This courtship is a reaction-chain type of courtship like that of the stickleback: male and female perform alternating activities and each requires a specific response from the other before proceeding. As in the stickleback, this procedure seems to ensure correct identification.

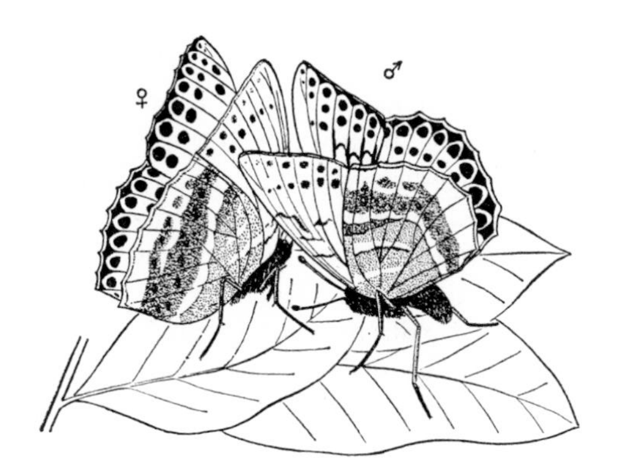

Figure 16. Ground courtship in the fritillary *Argynnis paphia*. The male begins his bow; this ends in his clasping the female's body between his wings.
(After Magnus [169].)

In sticklebacks, appearance, postures and movements provide the significant cues; in the silver-washed fritillary there are two significant sets of stimuli; visual ones, operating over a distance to initiate courtship; chemical ones, operating at close quarters. Hence the fritillary, like the cockroach, uses for identification the very sensitive chemical sense organs, typical of many arthropods.

The importance of chemical stimuli is presumed, rather than demonstrated, in much of the fritillary's courtship. However, a very recent study by Brower *et al.* of the displays of the Queen butterfly *Danaus gilippus berenica* [46], gives a clearer picture. Males of this species possess 'hair-pencils' which they extrude from the end of the abdomen. These consist of bundles of scales, elongated into hollow hairs, each continuous with a secretory cell. They produce a scent which smells fragrant to us; and during both aerial and ground courtship, the male bobs up and down above the female, sweeping these pencils over her antennae. Copulation is never successful in the absence of this activity.

Scent stimuli are usually highly specific in insects; visual stimuli initiating approaches may be less so, for occasional mistakes can be rectified. However, complex visual reactions are sometimes found where mistakes are likely to be common. The females of an African butterfly, *Hypolimnas misippus*, mimic butterflies of another species, *Danaus chrysippus*, which occurs in large numbers in the same area. The resemblance is close; both are a similar size and fly in the same way; both have orange-brown wings, although the *Danaus* butterflies have large areas of white on their hind-wings. Stride [259] showed that *H. misippus* males are attracted by orange-brown butterflies but avoid those which have large areas of black or white on their hind-wings. By this means they avoid, not only the *Danaus* butterflies, but also other males of their own species which have black wings. Non-specific clues are ignored; these insects, unlike the fritillary, ignore the form of movement.

Many other insects, besides butterflies, are solitary and have similar problems of meeting and identification. They may solve them in quite different ways. Grasshoppers (Saltatoria), for example, spend much of their time walking through long grass; they are unlikely to be guided to their mates by sight or smell. Instead they have complex and highly specific songs, produced by rubbing the hind-legs against the wings and perceived by tympanal organs on the legs. These songs not only attract, advertise and identify, as in birds, but also constitute the whole mating display. Above a certain temperature males sing spontaneously, sounding a number of notes followed by a pause. A receptive female answers with a softer call. Male and female then call alternately (the male with a somewhat altered song) while the male approaches the female. Courtship is yet another variant of the normal song and so is threat (male answering male). Perdeck has shown that the song is sexually stimulating to both sexes. Females who can hear the song of captive males of their own species become more receptive (even to the extent of mating with silent males of the wrong species). Likewise males become sexually aroused on hearing the song of captive females; they increase both their locomotion and their attempts at copulation [209].

Songs differ between species, especially species that share a habitat; and these differences are discriminated by females. Perdeck shewed that, of two species, *Chorthippus biguttulus* and *C. brunneus* which co-exist in the Netherlands, the females answer the songs of their own males only.

The salticid spiders

The salticids are free-running jumping spiders which stalk and leap upon their food. They too have problems of meeting; but they hunt on open ground and have good eyes, so they can use visual clues. However, they have also a less usual problem, that of distinguishing their mates from prey and, even more important, of being so distinguished by them. For spiders attack insects of roughly their own size and shape.

Drees [81] has analysed some of the characteristics which an object must possess before it is investigated by a salticid as a potential mate (figure 17). Size, movement, solidity and a broken outline are all important in attracting attention. These features are common to both prey and mates although, on the average, prey objects are smaller than mates. Males are smaller than females, however, and are in some danger of being attacked by them. Additional 'spider-like' stimuli are required to stimulate sexual reactions. There must be legs of about the right thickness and proportions, held at a typical spider-like angle to the body. Hence, again, as in *Hypolimnas*, there are very complex releasing stimuli for courtship, and these presumably ensure correct identification, even from a distance. This is essential, for a mistake could mean death.

Crane [65, 66, 67] gives detailed descriptions of many salticid displays. On sighting a female, a male begins his courtship with a slow and hesitant zig-zag approach. There are frequent pauses, during which the male displays. The exact form of the display varies

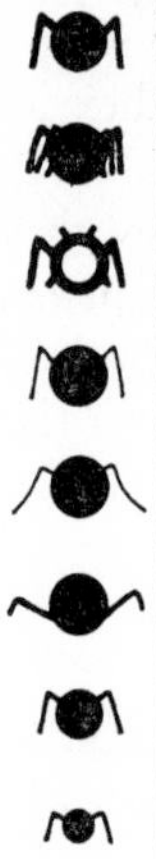

	n	C%
a	88	58
b	80	85
c	82	49
d	69	36
e	72	36
f	44	23
g	54	52
h	36	17

n = number of tests
C% = courtship responses

(*a*)

Days S		n	A%	C%
5		34	53	47
10		28	68	32
15		30	87	13
5		32	19	81
10		30	23	77
15		38	34	66

Days S = days starvation of males
n = number of males tested
A% = % attack responses
C% = % courtship responses

(*b*)

Figure 17. (*a*) Some of the models with which Drees induced courtship from satiated male salticid spiders. The models were moved: the proportion of responses is indicated. Besides body size, the position, number and thickness of the legs appears to be important as well as their angle of inclination.

(*b*) The same model may be attacked as prey or courted as a female by a hungry spider. The tendency to attack increases with the hunger of the male but decreases the more closely the model resembles a female spider. Attack and courtship reactions are shown for males at three different levels of hunger with two different models. (After Drees [81].)

with the species. *Ashtabula furcillata* (figure 18) holds the carapace high, waves the first legs up and down and turns the abdomen to one side. These movements show off the colour features of this species, namely, the darkened first legs and the striking abdomen which has a white stripe bounded by iridescence. *Corythelia fulgipedia* (figure 19) has a more complex display. The carapace is lower and it is progressively lowered during the display. The first three pairs of legs are stretched out and the third ones, together with the

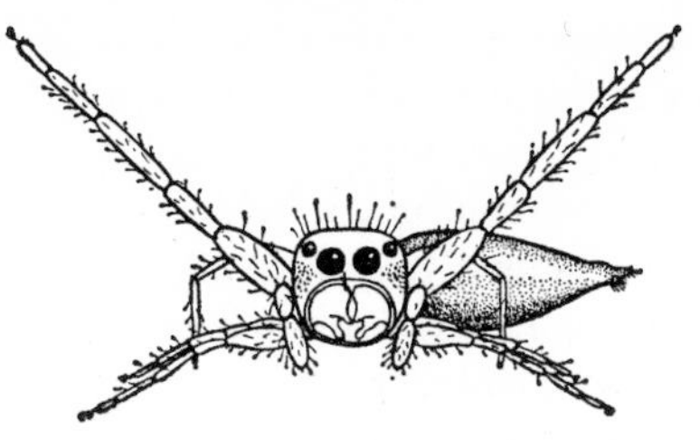

Figure 18. The courtship of *Ashtabula furcillata*. The male sidles back and forth, raising his front legs at a wide angle and waving them. (After Crane [66].)

palps and body are vibrated. The first three pairs of legs of this species are all fringed and have iridescent patches, while the third legs are especially elongated. The palps have white patches.

During the approach of the male, the female usually watches in a braced, high position. Sometimes she retreats, in which case the male usually follows. When she becomes quieter and more attentive, he speeds up his display, often standing still a short distance from her. Finally, if she crouches low, with her legs drawn in, he approaches rapidly, with his front legs held up and forward ready to deposit his spermatophore, before he makes a hasty retreat. Scent stimuli are probably perceived at close quarters, for male and female vibrate their palps and fore-legs, and these are both richly supplied with chemo-receptors. For some species scent may be more important than for others, since some salticids hunt mainly by sight, some mainly by scent; and they tend to court likewise.

There are obvious functional explanations of many features of this courtship. The reaction-chain pattern again appears as an insurance against mistakes; so do the highly specific visual (and probably

Figure 19. The courtship display of *Corythelia fulgipedia*. The male vibrates his third legs during or between spurts of progress towards a female.
(After Crane [65].)

also scent) cues. Moreover a male makes himself large and conspicuous, signalling his identity with startling colours and movements; and he never comes close to the female until she signals her willingness to mate with her crouching soliciting posture. The females of some species, for example *Corythelia fulgipedia*, give a reciprocal display before soliciting. This seems to happen where the spiders are particularly aggressive, and there may be danger of males attacking females.

There is also, in this group, some evidence bearing upon the origins of courtship. The behaviour observed suggests that there are conflicting internal states – an unusual feature of arthropod sexual behaviour. The male's sudden flights and hesitant approach indicate that he reacts to the female as both a possible mate and a possible predator. He seems simultaneously impelled to approach her and to escape from her. There is a remarkable resemblance between the salticid's approach behaviour, with its zig-zag course and intermittent displays, and that of many male birds which seem to fear their mates. And the female's soliciting posture also resembles that of some birds in being very distinct from the typical aggressive posture (salticids attack from a braced, high position). Hence factors promoting escape are minimized, and a male often rapidly approaches a soliciting female.

Many other spiders have the same problems. In some species, as in the completely unrelated mantids (Mantidae), males often *are* devoured during copulation. The orb-web spiders (Argiopidae) [42] have an unusual method of signalling their identity. The females are approached while in their webs, often at night; and the males drum on the suspension cord with a special rhythmic beat, quite unlike that produced by a struggling prey. The approach is hesitant and is interrupted by sudden flights whenever the female stirs. Once the pair have come together there is much stroking of the female with palps, feelers or legs until she turns on her back and allows the male to insert his spermatophore. In this case the female is required only to stay still, before the male will approach, but this in itself signals recognition, for she would immediately run up to an entangled victim.

The fiddler crabs

The fiddler crabs, *Uca*, also described by Crane [64, 68], are chosen as examples of that relatively rare phenomenon, arthropods with territorial behaviour. Although examples of arthropods with at least incipient territoriality are increasingly coming to light, it remains true that the majority lack it. In addition, the group contains many closely related species, each with a highly distinctive display; hence, like fruit-flies, fiddler crabs are of considerable interest from the evolutionary point of view.

Fiddler crabs live in large communities, often many species together, on a muddy or sandy shore in the Indo-Pacific or in America. Territories are held by both sexes, often throughout the year, and they are primarily used for food and shelter, each crab digging a burrow within its territory and retreating there when covered by the tide or on hot, sunny days. Nevertheless territories do appear to have some significance for courtship. Males tolerate the burrows of males of other species much closer to them (2–5 cm) than they do those of their own species (about 8 cm). Moreover, a courting male, while allowing young crabs, females and even males of other species upon his territory, drives away males of his own species. A courting male also builds a shelter or dome over the entrance to his burrow, and this may attract the attention of females.

Fiddler crabs are markedly sexually dimorphic. This is perhaps surprising in a species where both sexes are territorial, but in fact females are not very aggressive and never display. The most striking feature of the male is his one enormously enlarged claw which is waved in a highly characteristic manner both in courtship and in threat. Males are also very conspicuous, with the carapace (or back shield) often pure white, and the claw brilliantly coloured. Striking combinations of orange, yellow, lilac, salmon-pink and white occur on the claw and also on the front aspects of some of the walking legs. Females possess some of these colourings but are, in general, much duller. Also the colours of a mature male brighten each day on exposure to the sun, and he does not display with full vigour until this process is complete. Sometimes this takes an hour.

The display of the fiddler crab consists of a rhythmic waving of the enlarged claw; this is accompanied by body movements, so that it has the appearance of a dance. The movement may be simple or complex, slow or fast, according to species. It is always very stereotyped and specific. An example of a simple courtship is that of the Philippine species, *Uca zamboangana* (figure 20). A series of vertical waves of the claw is given at a high tempo, the carapace

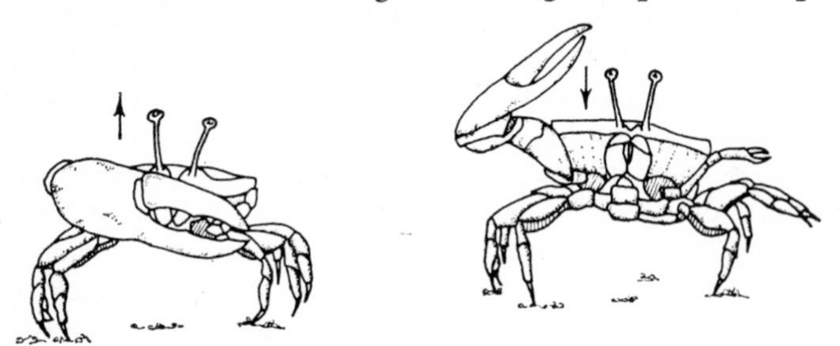

Figure **20. The display of *Uca zamboangana*. The enlarged claw is moved vertically up and down. The diagrams show the two extreme positions. (After Warren, from Crane [68].)**

being raised and lowered with each wave. The male displays on his own territory but later approaches a female on hers, seizes her, taps or strokes her carapace with his walking legs and attempts copulation. This takes place, on the surface, near the female's burrow.

A more complex courtship is found in *Uca pugnax rapax* (figure 21), which lives on tidal salt marshes from the Gulf of Mexico to Rio de Janeiro. The body is raised as high as possible and the claw is

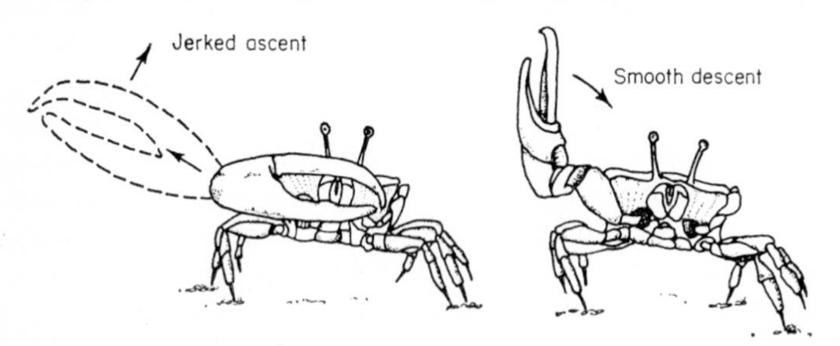

Figure 21. The display of *Uca pugnax rapax*. The enlarged claw moves outwards and upwards in three jerks, but descends smoothly. (After Warren, from Crane [68].)

brought up in a series of three jerks, moving laterally as well as vertically, held for a brief instant in the top position and then returned in a smooth downstroke. The pincers of the claw are opened in each display and one or two walking legs may be flicked outwards at the peak. Sometimes the male takes a few steps sideways while displaying. He usually gives about twenty displays in one series, each lasting about six seconds, which is slow for a fiddler crab. Unlike *U. zamboangana*, he does not approach a female, but waits for her to come to him. When she does so, he courts even more vigorously and leads her to her own burrow, which he then enters. A willing female follows him down and copulation occurs below ground.

In general, fiddler crab courtship can be divided into two main types, illustrated by these two examples. The more primitive Indo-Pacific and neotropical crabs all have the simple vertical wave and surface copulation of *U. zamboangana*, while the American crabs have the more complex, lateral wave and copulation within the burrow, as seen in *U. pugnax rapax*. The more advanced forms of both groups show increased complexity both of rhythm and form; the American crabs, in particular, tend to add other movements such as steps, kicks, curtseys and bobs, particularly as the female approaches. All the displays are very conspicuous and very distinctive. Crane states that she can distinguish even closely related species from a distance by the form and rhythm of waving [68].

There has been some controversy about whether these displays should be regarded as courtship at all. They exactly resemble threats given to other males, and there is no indication that they are directed at any particular crab, male or female, nor that the female observes them at all. It has been suggested that the so-called courtship is merely a normal threat display which happens to be followed by an approach to or by a female. However, the displays are always intensified as a female draws near (or shows interest) and, where there are extra elements, like curtseys and bobs, these appear only in the presence of a female. It seems likely that the displays originated as threats (to males) but during evolution came to have a stimulatory effect upon females and are now specifically evoked by the female's presence. If so, they are examples of courtship.

The courtship of these crabs, like that of so many arthropods, seems to prevent cross-mating – a function of particular importance where many species congregate together. Males of each species give a unique display which their females presumably discriminate from others. Crane's observations on *Uca pugnax pugnax* support this idea. This subspecies occurs in pure colonies, and Crane reports unusual variation in its display; it is also briefer, less elaborate and much less conspicuous than that of near relatives living in mixed colonies [68]. In other words, variable and less distinctive displays can be tolerated in pure colonies; in mixed ones they would lead to confusion with other species.

Fiddler crab territories are evidently related to feeding and protection. But at mating time they are used as display posts with each male a little separate from others of the same species. In this sense they resemble the mating territories of lek birds, although they provide even less isolation. Contact between the mates is brief and it is probably not possible for prolonged individual displays to develop. This may explain why no chain-type displays have developed for identification as in salticids and butterflies.

Survey

When it exists, arthropod courtship may be surprisingly complex. Usually the displays are very stereotyped and less variable, both in form and occurrence, than those of vertebrates. Perhaps because of this, it is much more difficult to recognize arthropod displays as modified forms of other behaviour patterns. Among vertebrates, duck displays look like modified preening movements, whitefish 'sailing' seems to be inhibited swimming and so on. The waving of a fiddler crab's claw or the vibration of a fruit-fly's wing is much less easy to interpret in this fashion.

In fact, evolutionary origins are, in general, obscure. In vertebrates evidence of behavioural conflicts in the mating situation can be obtained even from simple observations. But the attacks, the threats, the precipitant flights, so apparent in courting fish and birds, are often completely absent in arthropods. This is unlikely to be an error of observation or interpretation, for in the few cases where there is good reason to believe that males *are* ambivalent in their attitude to females (for example in the salticids), this is immediately apparent from their behaviour, which shows remarkable parallelism to that of birds in similar circumstances. By contrast, most arthropod males approach females without any apparent hesitation or fear and they rarely flee from them, for most are not attacked, in any normal sense of the word. Even the repelling activities of coy females do not inflict injury. Nor, with rare exceptions, do arthropod males show any tendency to attack their females. Where males fight, as among fiddler crabs, this does not appear to extend to females. Hence it is not easy to claim that arthropod displays arose from conflict behaviour. This problem will be discussed in a later chapter.

By contrast, functional explanations of arthropod displays are fairly obvious. There are three very common functions: to stimulate and manœuvre 'coy' females for copulation; to facilitate the meeting of solitary animals; and to facilitate identification where there is danger of cross-mating. Some of the additional functions found in fish and birds are lacking. There is no need for close synchrony of physiological processes as in fish simultaneous spawners, nor indeed even for such physiological timing as is required by birds with internal insemination; for most arthropod females can store sperm over long periods. There is usually little need for appeasement or synchronizing moods, there are no nest-site displays and little luring. Courting males in communal groups do suffer from interference, but they usually repel other males rather than lure females away.

There is good evidence for the manœuvring and stimulating functions of much of arthropod courtship. Manœuvring the female into a useful position is well illustrated by the cockroach feeding device and by the fritillary's aerial display, which usually succeeds in bringing the female to the ground. Stimulating qualities have been demonstrated for fruit-fly wing displays and grasshopper songs among others. Moreover there is evidence (from their behaviour) that the virgins of many arthropod species *are* coy and require such stimulation before they will mate.

The problem of the meeting of mates may be very acute for arthropods which are solitary and wide-ranging in their habits, for all the species concerned lack acute distance vision. Some of the devices employed have been described: scents, sounds and 'flickering' visual stimuli are common. Sometimes, as in vertebrates, there are communal gatherings of males for communal advertisement. Burmese and Siamese fireflies congregate and flash in unison [287], and crickets (Gryllidae) sing in chorus.

For the avoidance of cross-mating, releasing stimuli must differ significantly from similar stimuli provided by other species living in the same area, and these differences must be discriminated by potential mates. Many types of stimuli are used; grasshoppers discriminate sound differences (usually on the basis of amplitude variations producing a typical chirp rhythm rather than by pitch) [109, 157, 209, 283], fireflies detect differences in the grouping of flashes (a single flash versus a triple one in the N. American *Photuris* species for example) [16, 168], moths and cockroaches discriminate odours and tastes; salticids, butterflies and fiddler crabs, complex visual patterns and movements. This is perhaps the most important function; many arthropod displays appear to be little more than devices to direct the mate's attention towards the relevant releasing stimuli.

Part Two: Evolution

4: Introduction

The theory of evolution by natural selection is now so well authenticated as to leave only a few stubborn doubters. There is abundant evidence that populations of living organisms have slowly changed with time, usually in parallel with changing circumstances.

This process has been possible because separate, interbreeding populations share a pool of genes which is distinct from that of any other group of animals. The genes are to some extent variable so that individual members of a population are not identical but vary around a mean which represents the type best fitted to the particular circumstances. Extreme forms (and the genes which determine them) are eliminated because they cannot compete with fitter organisms. This is natural selection which, under constant conditions, tends to keep a population stable. But, when circumstances change, slightly abnormal individuals may become fitter than normal ones. So the gene pool will slowly shift to produce a new norm. Hence this natural selection brings about evolution.

Every organism develops according to a pattern determined by its genes and their reactions with the environment. Every aspect of that organism depends to some extent upon genes and can be modified by them. And this means that its species can evolve. Behaviour is no exception, although it may sometimes seem so because so much of it depends upon circumstances and experience. Yet most animal species have unique behavioural repertoires; a cat may vary its hunting behaviour considerably but the result is always typically cat-like rather than dog-like. This can only partly be explained by the different conditions and upbringing of kittens and puppies; the rest must be related to the genetical differences between the species. In fact, animal behaviour depends upon anatomy, physiology, sense organs and nervous organization, and all these are modifiable genetically.

The relative importance of genetical differences in determining behavioural differences can most easily be assessed by studying relationships. Closely related species (whose gene pools have diverged but little) are likely to resemble one another more closely in genetically determined characters than are more distantly related ones. And differences between the displays of closely related species, besides being small, should resemble those found between individuals within each species. Courtship displays, in general, confirm these predictions.

Hence this book will not ask the question, did courtship evolve? but only how did it do so? The next three chapters deal with three aspects of this problem. The first is about selective forces – the power behind the evolutionary process. The second traces the possible course of the evolution of displays. The third discusses the genetical background to all this.

5: Natural Selection and Sexual Selection

Under what circumstances does an animal become fitter because it displays? Evolution is a process of change; it occurs when new conditions create new needs and hence give advantage to new sorts of individual. In the evolution of courtship this means individuals whose behaviour and appearance before mating is stimulating and attractive. Their advantages must be such as to make for better survival or more efficient reproduction (that is, they must be fitter). What kind of needs could such individuals fulfil?

Reproductive efficiency depends upon many factors. It is not enough to mate and produce a large number of offspring: the offspring must be healthy, and likely to survive and reproduce well in their turn. For this, they must often be produced at the right place (with protection and food available), at the right time (to coincide with the maximum food supply) and in the right numbers (to ensure that all are reasonably fed). The last point is especially important for large animals, where a smaller batch of young may grow into healthy adults, whereas a larger one would become half-starved weaklings, easy targets for predators and disease. Even more important is that the offspring should be the result of matings within the species, for hybrids are often inferior or infertile. These are the needs. If circumstances arise in which exceptionally stimulating individuals are better able to realize one or another of them, these individuals will be fitter. The resulting process will be natural selection, and the population of displaying individuals will be reproductively fitter in the given circumstances.

Darwin envisaged only one type of circumstance for the evolution of courtship, and one type of advantage [77]. The circumstance is that where males compete for females, the advantage (won by the more attractive males) is to mate with more females or to mate more quickly. This process is a special case of natural selection. The winning of mates is an essential part of the reproductive process and, if displaying males are more efficient at this in a competitive environment, then they are, by definition, fitter individuals. In the same way, to attract customers is part of the business of shopkeeping and, if advertising shops are more efficient at this, in a crowded community, then they may survive where others fail.

Darwin recognized this point. Nevertheless he distinguished this type of selection and gave it another name – sexual selection – because he believed its end-result to be different from that achieved by natural selection. The end-result of natural selection is a population which is fitter than its predecessor. For example, the selection of individuals which avoid cross-mating, where similar species live together, results in a population which is fitter because it will leave fewer hybrids. Darwin stated that sexual selection does not have this result. He held that the whole advantage of displays favoured by sexual selection lies in the satisfaction of female whims. Once evolved, such displays may have no influence upon reproductive success, nor even, paradoxically, upon winning a mate. This is because, where competitive interactions between animals are concerned, new or unusual individuals can create the demands which they satisfy. But their advantage is lost once they become the norm. Given a choice between stimulating and plain males, a female may mate with the former, but once all have become stimulating she may mate with them no better than she did with the original population of plain ones.

This point is well illustrated by observations on the fruit-fly *Drosophila melanogaster*. Males of a mutant form with vestigial wings fail to win females in competition with normal males because they cannot perform the stimulating wing display [71, 224]. However, pure cultures of vestigial insects mate and reproduce just as efficiently as do normal ones. It seems that normal insects have gained nothing from their evolution of a display (for they do not normally compete with vestigial flies). In other words, this population, which has evolved a display, is no fitter than its predecessor, even although the competition for mates remains.

It is now certain, however, that competitive situations are not solely responsible for the evolution of courtship. We first consider other circumstances and other needs, calling their outcome natural selection. Then it will be easier to assess the significance of sexual selection as Darwin envisaged it.

The case for natural selection

We have to guess the nature of the selective forces which bring a display into being. This is done by considering its present functions. Where there is only one important function, this is easy. Where there are many, only one may be significant in making the animal fitter. Alternatively, one may have preceded another in time, so that an existing display is put to a new use in new circumstances.

One of the most obvious functions of many displays is that of bringing mates together at the right time. Many animals breed seasonally and have physiological mechanisms which cause ova and sperm to ripen at the appropriate time. Displays may then help ripe males and females to come together quickly so that time is not lost in searching. Many conspicuous displays performed alone, in the absence of a mate, have this function. They include territory patrols, display flights, swimming movements and postures, songs (of birds, grasshoppers and frogs), scent dissemination (by moths and mammals), light signals and other devices. Sometimes they are performed only by the males, sometimes (among grasshoppers and fireflies) they are employed in call and answer sequences between the mates, occasionally (in moths) they are used by females to guide males to them.

A strong case can be made for the natural selection of all these displays. Circumstances which would favour individuals performing attention-catching activities might arise in populations which increase their range, and so make it more difficult for individuals to find one another. This applies particularly to invertebrate animals, which are small and often unable to see well over long distances. The development of territorial behaviour might create a similar need, especially in migratory animals where the later-arriving females have to find males on their territories.

A second function, that of ensuring that mating occurs in the right place, is often served by these same displays. This is obviously so for territorial animals but non-territorial ones too, such as white-fish, display only in specific localities. In other species there are luring displays and nest-site displays, enticing the partner to the nest or spawning place. This function may be a very critical one when the welfare of the young makes special demands. Bitterling *Rhodeus amarus* eggs must be laid in mussels *Anodonta*, stickleback eggs in special nests which can be ventilated, fighting fish *Betta splendens* eggs in bubble nests on the surface of the water. Hence, wherever these requirements exist and wherever egg-laying and fertilization coincide, that is to say, where both mates must come together in the appropriate place, then individuals whose behaviour aids this process are likely to be more successful reproductively. Clearly physiological changes in the young, leading to these special requirements, will often create such a situation, but the young's requirements cannot become very specific until the behaviour which makes them possible has itself evolved. This is one of many examples where the evolution of two interrelated characteristics must proceed hand-in-hand.

A third function appears to be that of synchronizing physiological processes and moods in male and female. Courtship displays may trigger-off processes concerned with the release of ova and sperm. They may bring male and female together into mating mood, overcoming any barriers which may exist between them (figure 22), or, over a longer period, they may be concerned with the maturation of female gonads so that their ripening coincides with the availability of the male.

The triggering function is most apparent in simultaneous spawners, where it is important for both partners to be brought simultaneously to the pitch of excitement necessary for the shedding of gametes. The synchronizing of moods is important in a number of situations. The male of territorial species is initially aggressive towards the female and cannot mate while he is still inclined to attack; among the passerine birds, females tend to attack males when they approach to mount, and the males are consequently timid; either male or female spiders may attack the other as prey.

Long-term effects upon the maturation of the ovaries of some

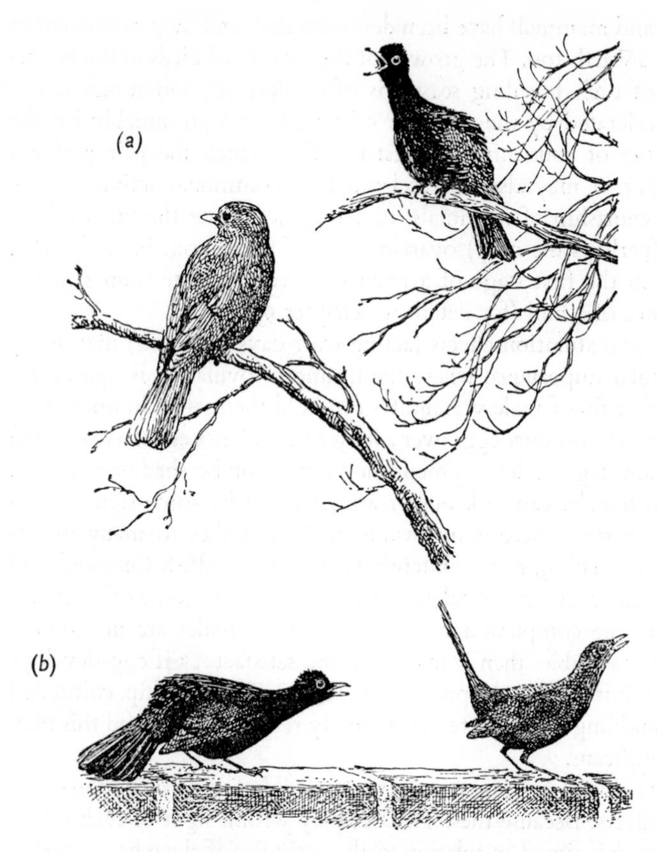

Figure 22. Courtship as a means of establishing the pair-bond in the blackbird *Turdus merula*.

(*a*) Early in the season a male courts his own (or other) females for long periods. Both birds seem 'nervous'; the male never approaches a female directly and the female, although impassive, flicks her wings and may hop away. These displays may lead to the establishment of a pair-bond after which (*b*) courtship is always brief. It is initiated by the soliciting of the female which stimulates the male to approach in intense courtship and mount immediately.

(From Snow [252].)

birds and mammals have been demonstrated, and they may occur in other animals too. The growth of the ovaries of birds at the beginning of each breeding season is often delayed; sometimes it can be accelerated by the presence of a male and presumably by the activities of courtship and nest-building which the pair perform together. It may also be accelerated by communal activities. The male courtship of mammals, too, may accelerate the onset of the first (perhaps the only) ovarian cycle of the season; in cattle, *Bos*, cows in the first stage of a recurrent cycle ovulate sooner in the presence of a bull (for details see chapter 9).

These maturation effects (acting over days or weeks) may be of particular importance where the timing of ovulation is significant. Probably few female vertebrates can hold themselves in immediate readiness to produce eggs over a long period. Once eggs have reached a certain stage of development they must soon be shed or resorbed. When females can seek out or attract males by some signal when they are ripe, there is no problem. This applies to many insects (moths, fireflies), fish (whitefish *Coregonus*, goldfish *Carassius*) and mammals. But when ovulation must be related to some other event, such as the completion of a nest, or when males are not always readily available, then it may be more satisfactory if egg-development is initiated by the presence of a male. The courtship, coitus and nest-building of birds are often closely related in time and this may be significant.

For colonial birds there is another possible significance of maturation effects. Because these birds display within sight of each other, ovulation is timed in relation to the activities of the whole colony, not just the pair. Within any one colony, all the eggs may be laid within a space of a few days. Patterson [207], after a study of this in black-headed gulls *Larus ridibundus*, suggested that it might cut down predation, for predators can cope only with a limited number of prey each day, even if they can store. Moreover, of the main predators of the black-headed gull (crows *Corvus*, foxes *Vulpes*, stoats and weasels (Mustelidae) and other gulls (Laridae)), only crows seem to store. So if, for example, a thousand eggs are produced over a period of a week, fewer are likely to be taken than if 250 are produced each week for a month.

These many different advantages conferred by synchronization are all such as can be derived from rapid mating or critically timed mating or ovulation. However, it is not always clear that some of the advantages suggested are really significant in improving fitness. Displays which prevent attack or escape, for instance, may make mating easier and quicker. But how often does this effect the welfare of the offspring? We cannot answer these questions without a full knowledge of the circumstances. It may be very important for many birds to catch the best of the food supply, or fit in an extra brood; it may be very significant for a stickleback to fill his nest with eggs of nearly equal age. Perhaps, too, the reduction of fighting increases fertility or reduces the risk of injury. Observations certainly suggest that delays due to interfering factors can be considerable. Snow [252] observed young first-year male blackbirds *Turdus merula*, which had been forced to contend fiercely against well-established males for territories and were very aggressive. They were also aggressive towards their females and consequently interfered with and delayed the latters' attempts at nest-building. Braun [41] reports similar findings in captive song-birds.

A fourth function of displays is that of preventing cross-mating. Very often courtship emphasizes specific stimuli which are distinct from those provided by other species. Animals which can limit their sexual reactions to such stimuli are likely to leave more (and more fertile) offspring. Sometimes these specific stimuli are part of the signals which guide mates to one another (as in fireflies, grasshoppers and birds); sometimes stimuli available only at close quarters are more important (as in the cockroach and some butterflies), sometimes the whole courtship seems to consist of a series of identity checks between male and female (as in the stickleback and the fritillary).

Courtship displays are not the only source of identification signals. Sometimes general features of appearance, scent or behaviour are sufficient. For example, different related species of the lizard *Sceloporus* tend to bob their heads each in a specific rhythm. This is not a display (they do it all the time) but females can identify males of their own species by means of it [123]. Very often, however, displays serve to attract attention to existing distinctive features by wafting scents or moving coloured parts; other displays are themselves unique (in movement or sound).

A point should be made regarding the relative value of this function to male and female respectively. It is sometimes argued that discrimination is most often a feature of female behaviour because females, at least of polygamous species, suffer most from mistaken matings [27]. Females may mate only once in a season while polygamous males mate often, hence a larger proportion of a female's total offspring will be affected by a mistake. However, discrimination by males is, in fact, very common and not only among monogamous forms in which both sexes suffer equally. Male fruit-flies frequently turn away from foreign females as soon as they have tapped them, so do male cockroaches. But when males discriminate, females usually do so too, and the latter seem to be more absolute in this respect. Males of many fruit-fly species court some foreign females (rejecting others), while females rarely accept any but males of their own species. Probably female discrimination is the more important in preventing cross-mating, while male discrimination is mainly of value in preventing the waste involved in courting an unresponsive object, especially in a competitive situation.

The circumstances likely to make this function of selective significance are those where an overlap occurs between two populations of animals which might be confused with one another. Then 'coy' females will be at an advantage, for they are less likely to make mistakes, especially if they require extra stimulation of a specific type. 'Coyness' in its turn gives advantage to male displays, especially in a competitive situation, and the more these meet female requirements, the more effective they will be.

A fifth function is that of deterrence, of keeping other males at a distance. Sometimes it appears that this function is served by separate threat displays which are distinct from courtship. However, further consideration suggests that many threats occurring in the breeding season also operate as courtship. Threat was defined earlier as a social signal which tends to cause withdrawal without injury on the part of an adversary. In the breeding season, threat commonly occurs in encounters between males; in disputes over territories or over females, it seems to cut down fighting unless the dispute is severe. Identical displays are also directed at females, particularly during early encounters, when females appear to provoke reactions normally given to male intruders. Where immature females are concerned, these displays function as threats; they cause withdrawal. But ripe females do not withdraw; on the contrary they appear to be attracted and stimulated not only by threats but even by attacks. Domestic cocks (*Gallus*) which induce hens to crouch most frequently are those which are the most aggressive and which chase, waltz and strut (all threats) before them most frequently [291]. And female bullheads (*Cottus gobio*) respond sexually (by entering the

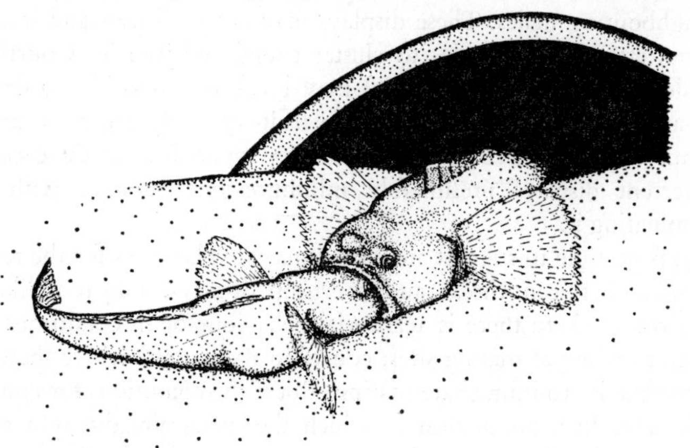

Figure 23. The courtship bite of the male river bullhead *Cottus gobio*. (From Morris [192].)

nest) to a bite from the male which in no way differs from that administered to a rival or to any other intruder near the nest [192] (figure 23). Even when threats and fights occur exclusively between males (as in swordtails, fighting fish, blackcocks and deer), they too may stimulate onlooking females. Hence these displays have a dual function: they operate as threats to most members of the species but to ripe females they operate as courtship and must be treated as such. Indeed in some species (fiddler crabs, domestic fowl and some gulls) the greater part of courtship is of this nature. And where this is so, a male, using a single display, is able to court a female and ward off rival neighbours simultaneously.

The advantages of threats in territorial species include all the advantages of breeding territories in general: seclusion for mating, for protection and the assurance of a nearby food supply for the young, among others. The males of non-territorial species may drive rivals from good breeding places but, most often, the chief value of deterrence lies in the prevention of interference in courtship and mating. This is particularly important for gregarious or social animals whether they hold territories or not. For here (in sparrows, lek birds, fruit-flies and fiddler crabs) the displays between male and female which immediately precede mating are seen by many neighbouring males. These displays may not be threats and sometimes attract other males to cluster round and join in. Courting males (and sometimes females) may break off courtship to drive away such intruders, and a female fruit-fly rarely accepts a male while a group of courters is clustering round her. In these cases deterrent displays presumably minimize this nuisance without eliminating it.

It is possible to suggest an evolutionary explanation for the relationship between threat and courtship displays. Threat is likely to evolve where there is territorial behaviour or in conditions of overcrowding at mating sites; courtship will evolve where there is some barrier to immediate mating. But if the conditions for courtship arise in a population in which the males already fight and threaten, then it will be advantageous for ripe females to respond sexually to threat displays, for these signal the presence of a mature male in a good breeding place (this may be the situation in most territorial species). The reverse may not be true. Already existing courtship displays signal the presence of a female and it will be advantageous for males to approach, rather than avoid them, particularly in polygamous species. Then distinct threat displays may arise in courting males to minimize this nuisance but these will not normally acquire a courtship function (this may be the situation in the fruit-flies).

This account of the functions of courtship is not exhaustive. Many minor ones exist, such as positioning females for easier coitus, as in the cockroach, or luring them away for greater privacy, as in the guppy. The value of these must be assessed individually in relation to the mating and reproductive problems of each animal concerned.

The case for sexual selection

On sexual selection we consider three examples. The first is the peacock *Pavo cristatus* [218]. Polygamous males gather in an arena at mating time, where they display to any female which approaches. The tail is spread, the plumes are rattled, and the wings are quivered behind the tail fins. Hence, in addition to the arresting sight of the tail, there is a very effective noise to set it off. The display is directed mainly at the female, who, if she is aroused, runs round to the front of the male whenever he turns his back; and eventually squats.

Males display simultaneously in the arena, and hence compete directly. A female can compare a number of displays and choose the one which stimulates her most. Presumably the most attractive males will fertilize the most females and, if attractiveness carries no attendant disadvantages, these animals will be fitter. If the largest, most elaborately patterned and the best displayed tails do indeed confer this sort of advantage, we need look no further to explain their evolution. Moreover, any new exaggeration, however fantastic, if it provides added attraction, will be similarly selected. There is no evidence that the tail of the peacock is of any value outside the mating season; females survive well enough without it.

Peacocks, then, appear to be an admirable illustration, not only of the conditions suggested by Darwin, but also of the predicted outcome in terms of displays and display features. They are exaggerated to an apparently unnecessary degree and they seem to serve no function other than helping males to score over males. Only one recent doubt has been raised regarding the assumed attraction of bright colours for females. Dewar reports that albino cocks (figure 24) in the Lahore zoo are preferred to normal, brightly coloured ones, because they display more vigorously and more persistently [79]. Hence it seems that these females are choosing vigour rather than brilliance. However, this argument is really a red herring. Albinos have not been selected in the wild, perhaps because their conspicuousness makes them easy victims for predators. Hence females are not normally offered this choice. They have to choose from among coloured males, varying both in vigour and colour contrasts.

A slightly different type of competitive mating situation is that found in the fruit-fly *Drosophila melanogaster*. At the food source, where they emerge from their pupae, virgin females are surrounded

Figure 24. An albino peacock in display.

by males which begin to court them persistently as soon as the cuticle has hardened. It cannot be asserted that females actively choose the best courters, but there is little doubt that they do, in fact, mate with them. There are several mutants of this species which have a slightly less vigorous wing display ('yellow' is an example [24]); these are always less successful in competition with normal males. This parallels the similar failure of vestigial males (already quoted), which lack a wing display altogether.

In this species females do not respond to one of a number of simultaneously displaying males. They are often mobbed by many males courting at once, but they rarely accept under these circumstances. The most successful males are those which can get a female on her own and then fertilize her quickly before the opportunity is lost. Hence effective repelling displays (to drive other males away) are selected, as well as stimulating ones, and usually these are distinct. The stimulating displays promote a very rapid response, but (because there is no direct comparison of one male with another), they are unlikely to become as exaggerated as the peacock's tail.

A third example is a deer [75]. The stag of the red deer *Cervus elaphus*, a native of Britain, develops antlers each year for the rutting season. These increase in size up to about the twelfth year, and a complete set of horns (called a royal) is held from about the eighth to the eleventh year; they are very cumbersome, with a great span and many horns. Males compete and fight one another for the ownership of harems; any stag may challenge another already possessing a harem, but he usually does not dare to do so if the possessor is larger or has larger antlers. Fights consist of clashing heads and locking antlers, after which heads are twisted in an attempt to deal a blow on the flank which can be fatal. Victorious males take over the females of the loser.

In this case the antlers and their display are used for fighting and threat. The competition for females is a direct one between the males; there is no wooing of female whims. Indeed, the females are for the most part herded passively and seem to be indifferent to the process. Only a few (those already on heat) actively seek out males. In this sense this example differs from other cases of sexual selection.

Nevertheless antlers clearly confer reproductive fitness; those with better heads can serve more females. It is likely, too, that females are stimulated by the sight of a good head (as well as by the fighting and bellowing) to come on heat more quickly (hence the threats function as courtship too). This too is to the advantage of the big stags, for they wander away in November, leaving any females who have a late oestrus to be fertilized by the smaller unsuccessful stags.

Like peacock's tails, stag antlers appear to be excessive in their development. They have no other function; hinds do not possess them, nor do stags outside the breeding season. Moreover they constitute a considerable drain on the stag's metabolism, requiring large quantities of calcium salts and phosphorus; they may even be a liability when the velvet or covering skin is present, as damage to this can result in copious bleeding.

In deer, as in peacocks, some observations have thrown doubts on the fitness conferred by the sexually selected characters (in this case large antlers). Mutant forms lacking antlers (called hummels), contrary to expectation, collect and maintain large harems and hold their own in a fight [75]. This is partly due to the fact that they are very large (perhaps because of their freedom from the strain of growing antlers). However, again this is a red herring. This mutant fails to spread in the population, for some other reason, and does not normally take part in the competition. Another doubt comes from observations that stags with immature antlers are sometimes victorious in fights because they can twist and jab better and are less easily entangled. But this must be weighed against the superior intimidating powers of large antlers; the very big stags are rarely challenged. As with courtship features, excessive development may be promoted only when they compete to influence other animals.

These are all situations where males are polygamous and where animals collect together for mating. In other words, they are situations most favourable to Darwin's case. What of others in which males mate in isolation, perhaps on a territory, perhaps (in wide-ranging species) wherever they happen to meet? Here direct competition between males is impossible. And what of monogamous species where each male can win only one female, however attractive he may be?

Darwin's hypothesis regarding the evolution of the displays of some of these difficult cases is worth quoting even though, for most of them, it is almost certainly mistaken [77]. He believed that the females of monogamous species still choose to mate with the most attractive males (if they can), although this gives the latter no advantage; the plain males get females too. But Darwin suggested that the more attractive males might also be more vigorous, and hence leave more offspring for this reason. Maynard Smith [187] illustrates this idea from a human example. He points out that a preference of women for red-headed men in a monogamous human community in which all males married, would not lead to the selection of red-heads; these would merely marry first. However, red-heads who were also more healthy and more fertile would certainly spread in the population. Darwin believed that females could promote a link between general vigour and a characteristic favoured by them, by exercising a double-weighted choice, so to speak. Like the peahens, discussed earlier, they might prefer both vigorous and ornamented males. In addition, he supposed that the females able to exercise the widest choice would be the most vigorous of their sex. This could apply especially to seasonally breeding species, of which the most vigorous females might mature first. Hence, probably, the most successful couples in each season would be vigorous females mated with vigorous ornamented males and these would, by virtue of their vigour, leave many offspring to inherit both characteristics.

This is an ingenious theory and one which is theoretically plausible, even though the selective pressures are again likely to be small. Its chief drawback is that it equates several types of vigour – general vigour, vigour in display, sexual precocity and high fertility – which are not necessarily related. Admittedly, many of the mutant and inbred male fruit-flies which are for various reasons less fit (in the survival sense) also have a less vigorous courtship [24, 71, 92, 188, 222, 224]. Here, then (in a polygamous animal), some sexual selection of this type must occur. But there is no evidence that it happens generally. Albino peacocks provide an example of vigorous courters of low fitness.

From these examples it is clear that sexual selection does indeed occur in polygamous and perhaps in some monogamous species too. But is it ever the only factor in evolution? Do displays ever evolve with this function and this alone? A consideration of the circumstances which promote sexual selection throws some light on this question. With rare exceptions (provided by harem-holding animals like deer), an essential condition is that females should choose. This does not necessarily imply conscious choice, merely that they require to be stimulated and that some males stimulate them better than others. Not all females choose. Where a male house-fly jumps on to the back of a female, there is probably no choice. And there is little sign of it in some mammals like sheep and cattle.

In other words, males compete, and sexual selection occurs, mainly when females are coy and initially unwilling to mate. Female coyness is often related to reproductive needs: the need to avoid the wrong mate, to ovulate at some specific time, or to be guided to mate or breeding place. As discussed above, coy females, requiring specific stimulation, are at an advantage when such needs exist; so are displaying males which can help to satisfy them. But in certain circumstances, particularly where animals gather for mating, they will compete to do so. They will compete to guide, to provide rapid or specific stimulation, or to advertise a good breeding place. Indeed they are certain to do so, if (by mating sooner or more often) they gain reproductive advantage thereby. Hence, contrary to Darwin's prediction, displays which originally gave competitive advantage often have other functions and, in consequence, improve the reproductive fitness of the population.

Occasionally, while mating difficulties exist, they may not be important, and it may make no difference to reproductive success for them to be overcome sooner rather than later. This has been suggested for species in which there are moods of aggression or escape to be overcome. The displays of polygamous species could still give competitive advantage to males, by enabling them to mate sooner, but these would have no other significant value, and the process would be true sexual selection as Darwin envisaged it. However, we do not know whether there are any such cases. Even in fruit-flies, peacocks and deer, the fact that other functions are not obvious does not mean that they do not exist. The displays of the fruit-fly may help females to discriminate against wrong males (although they also do this by means of chemical stimuli); the displays of peacock and deer may accelerate and synchronize ovulation.

Hence sexual selection, in Darwin's terms, may not exist. Nevertheless, competition between males often plays a considerable, and sometimes a dominant role in the evolution of courtship displays. Where this is so, the displays will tend to have the characteristics which Darwin suggested: they will be elaborate and exaggerated, vying with other displays to attract the attention of the female. While this may not be the only cause of complexity, it must be an important one. Whatever their function, competitive displays will usually exceed any normal stimulus. A nest-site display may be relatively simple when it occurs between two already paired birds, such as yellow-hammers; it may be highly elaborate when used competitively to attract females, as by bower birds. Antlers are small and compact in roe deer *Capreolus capreolus*, of which each male keeps only one or a few females and fights are relatively rare [75]; they have become enormous in the red deer which uses them

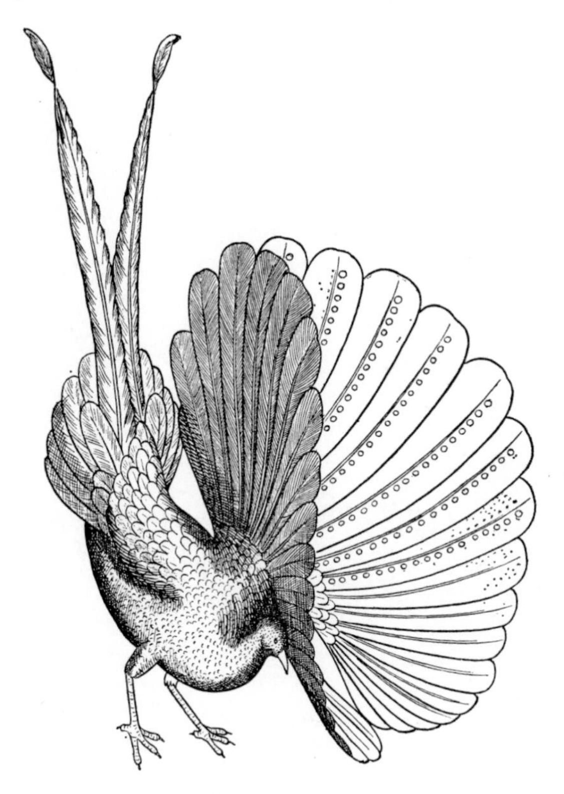

Figure 25. The display of the argus pheasant *Argus argusianus*. (After Darwin [77].)

competitively to defend enormous harems. We may conclude that the concept of sexual selection, if it has lost some of its unique importance, remains as an explanation of much that is otherwise inexplicable in courtship evolution. Who can imagine that the displays and adornments of the peacock, the argus pheasant (figure 25), the birds of paradise, even of the fiddler crabs are really necessary, in all their detail, to stimulate the female?

6: The Evolution of Courtship

It is one thing to discuss the advantages favouring animals which do conspicuous and attractive things before mating; it is quite another to ask how they have come to do them during evolution. When we look at courtship displays, many appear to have no direct relation to the process of mating. Some have. A male cockroach's display looks like the normal preliminary to mating; so does that of the whitefish. And certain manœuvrings, like following a female, or blocking her escape path, might also be regarded as part of the animal's normal repertoire of behaviour in pursuit of a goal (in this case mating). But many displays are interruptions to or even reversals of mating approaches: displaying yellow-hammers run away from their mates; pheasants and cockerels circle round them. Many displays feature organs, movements and postures which are not used in mating. How have some animals come to do such apparently irrelevant things?

Conflict situations

In the previous chapter it was suggested that courtship is most likely to occur when there is some difficulty about mating. Females may be coy and initially unwilling to mate, or there may be some additional reaction, in male or female, which prevents their coming together. The situation which has attracted the most attention is that where interfering or conflicting reactions may occur. In particular, males holding territories may attack or flee from intruding females as well as attempt to mate with them. These animals are said to be 'in conflict' and because of this tend to perform typical 'conflict activities' which could be the forerunners of displays. It is necessary to examine this suggestion more fully.

What does it mean to say that an animal is 'in conflict'? Psychologists mean that two incompatible activities are simultaneously provoked [43]. The animal must be responsive: it will not be in conflict if it is asleep or if it is indifferent to provoking stimuli. Psychologists can arrange such situations experimentally. They can place a hungry animal midway between two equally attractive dishes of food. Or they can place it midway between two objects from which it normally retreats with equal intensity. Or they can present it with an object which attracts it (like food) in a place which it has learned to avoid (perhaps because of electric shocks received there). In each case the strength of the approach or avoidance stimulated by the objects can be recorded.

The definition suggests a behavioural deadlock, and the behaviour of animals sometimes confirms this. It carries out neither of the stimulated reactions fully; sometimes it performs one or both of them partially, sometimes it does something else, sometimes it does nothing. Often it is restless.

In natural situations it is usually impossible to measure the strength of supposed opposing reactions. How then can a conflict be demonstrated? One way is to observe the behaviour of the animal as the situation changes. For example, a social bird, low in the hierarchy of its flock, may be hovering a short distance from a food bowl upon which a dominant bird is perched. If this bird is in conflict (between approaching to feed and retreating from the other bird) it will be expected to retreat whenever the dominant bird makes a movement in its direction, and approach closer if the dominant bird moves away.

Very often the strength of certain reactions is related to the closeness of special objects, and this can be previously tested. Attack by territorial males on an approaching intruder is very strong near the nest, but falls off as the territorial boundary is approached. Outside the boundary the same male is more likely to flee from the same enemy. Hence a confrontation at the boundary may represent a conflict situation, and this idea will be strengthened if oscillations between attack and escape are observed whenever the boundary is crossed one way or the other [270].

Evidence from postures, intention movements, fin, feather or fur

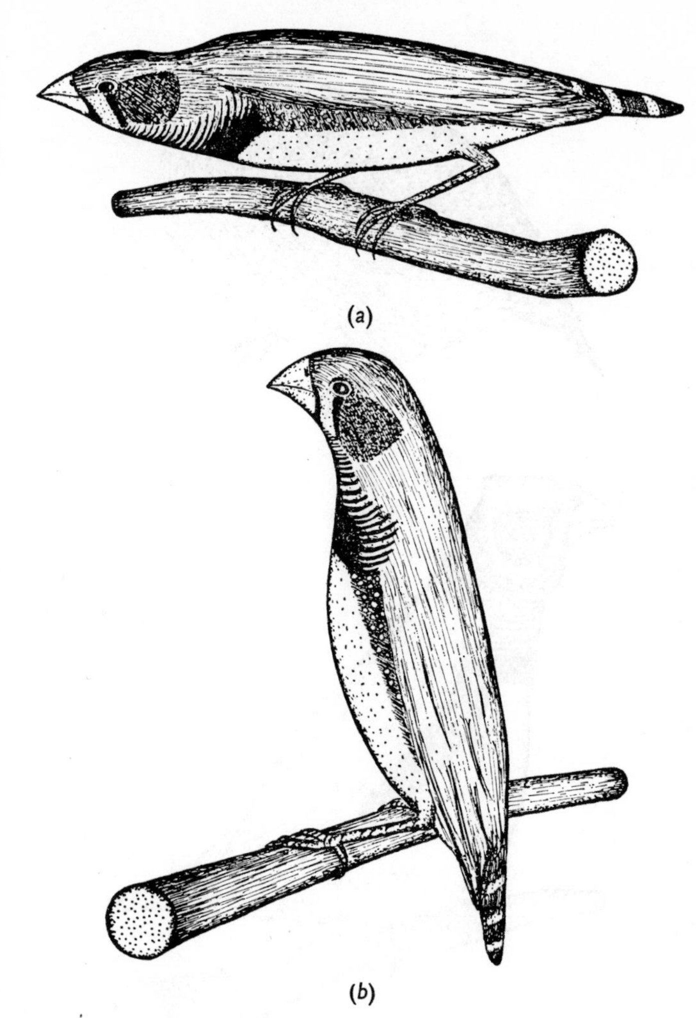

Figure 26. Distinctive postures in the zebra finch *Peophila guttata*. (From Morris [191].)

(*a*) The 'aggressive' horizontal posture: this is the posture of the charge, pecks being delivered at the body of the opponent. It also has threat function. (*b*) The 'frightened' sleeked vertical posture. Birds take off for flight from this position. Two evenly matched birds also beak-fence in this posture. In a fight, Morris claims that the more vertical the posture, the greater the probability of flight. (*c*) The

'submissive' fluffed posture. Feather fluffing is typical of rest. It is also typical of 'frightened' birds which cannot escape; such birds remain inactive and do not attempt to fight or flee. (*d*) The courtship posture. The upright body is here possibly a preliminary to mounting. The fluffed feathers distinguish this from the sleeked vertical posture of (*b*).

movements, is also helpful when their significance is well established. Very often the postures and preliminary movements of attacking, fleeing or mating are quite distinct and easily recognized. Feather fluffing or sleeking and crest raising may also have special significance for birds [194], while among mammals even facial expression may be used [176, 245] (figures 26 and 27). Finally some species of both mammals and birds have special calls or songs associated with specific acts. Hence an animal may indicate that a given act has been stimulated even although it does not perform it.

Even so, many of the claims about natural conflict situations are little more than guesses. It is necessary to examine these claims regarding mating situations with some caution. The strongest case

Figure 27. Expression of fear (*a*) and aggression (*b*) in dogs. In a frightened dog, the ears and the angle of the mouth are both pulled back; in an aggressive one, the mouth is opened, the upper lip raised and the snout and forehead wrinkled.
(From Lorenz [165].)

can be made at mating time for passerine birds like the yellow hammer. During engagement, the male is usually dominant to the female, supplanting her at food dishes. Yet, when he begins to show a tendency to court and mount, he appears suddenly subservient, giving way to her on all occasions. He may fluctuate from dominance to subservience, but observations suggest that his subservient behaviour is correlated with sexual activity [2]. When attempting to mate, he alternates between sexual approaches and sudden flights, while the female may attack him as he attempts to mount. Thus the male seems to be simultaneously stimulated to approach his female (for mounting) and to avoid her (to escape attack).

This suggestion is supported by the observation that sudden aggressive movements on the part of the female intensify flight reactions, while sexual reactions, nest-building and particularly soliciting (after which she never attacks) usually stimulate approach. In addition, the erect posture (associated with mounting) and feather fluffing (associated with thwarted escape) occur during courtship. On the other hand, it is difficult to believe that there is an exact balance between these opposing reactions during much of the yellow-hammer's courtship. In the fluffed run, for example, the male is predominantly escaping, albeit slowly, while the bill-raised run is a mating approach. Perhaps it is unnecessary to assume an exact balance, but rather that two opposing reactions interact and interfere with one another. This situation has been called FaM by Morris [195]. (The capital initials stand for fear and mating, the two principal opposing tendencies: the small a, for aggression, which may also be stimulated, though weakly.) A similar analysis by Hinde [112] provides a similar picture of the mating situation in chaffinches *Fringilla coelebs*. Among invertebrates, salticids and other spiders may also fit into this category.

A second type of situation has been called FAM. This is the situation of the territorial male on his first encounter with a female, especially where, as in fish and lek birds, he is going to mate almost immediately. It may occur also in non-territorial animals (like guppies and cockerels) which sometimes attack their females. Here, as the label FAM suggests, there is evidence of three separate reactions to females: attack, avoidance and mating. But before we can assume conflict we must ask three questions: are they equal, are they simultaneous and do they oppose one another?

Some quantitative evidence exists regarding the relationship between attack and sexual behaviour. In the three-spined stickleback Morris [195] counted the various types of behaviour occurring after certain typical courtship activities. Early in courtship (after zig-zags) males attacked on 70·4 per cent occasions, and gave sexual responses (leading to the nest) on 24·1 per cent occasions; late in courtship, however (after showing the nest entrance to the female), the corresponding figures were 5·3 and 93·7 per cent respectively. Meyerriecks reports similar findings during the flight displays of the green heron. Supplanting attacks made by males upon females were common after the early circle flight displays, less common after the crooked neck displays, and rare after the final flap flight display. At the same time, sexual tendencies, as indicated by snap and stretch displays, increased as the supplanting attacks decreased [189]. Finally Baerends studied the changing sets of black markings on courting guppies (figure 28). One set (pattern 2) was always associated with attacks, while patterns 1, 4 and 6 were associated with successive stages of courtship, 6 being always present at copulation. The aggressive pattern 2 appeared frequently with pattern 1, sometimes with 4, and rarely, if ever, with 6 [10].

All these investigations point to the same conclusion. Attack predominates in the early stages of courtship but later dies down and

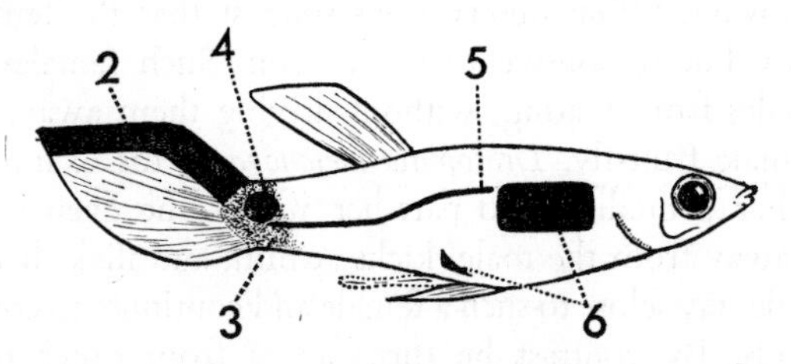

Figure 28. The black markings which may develop during courtship in males of the guppy *Lebistes reticulatus*. These markings come and go although more than one set of markings may be present at a time. Set I is not shown as it gives an overall darkness to the body. (After Baerends, Brouwer and Waterbolk [10].)

is superseded by mating attempts. The two reactions cannot be equally strong for more than a brief, intermediate period. Displays do not appear to be confined to such a point in time; they usually occur throughout the 'aggressive phase' and the 'sexual phase', albeit different ones in each case. Moreover it is questionable whether attack and mating reactions always oppose one another. Cats *Felis* [155] and cichlid fish [167], for example, appear to attack their females while mating with them. Only when sexual reactions involve an excursion away from the female (to the nest, by the stickleback, for example), can it be argued that they lead in opposite directions.

What then of escape? This has only rarely been studied quantitatively, but probably the tendency to flee from the female, should she move quickly towards the male, often remains high throughout courtship, and sometimes increases towards the end [10]. Escape conflicts both with attack and with mating; hence a conflict might exist between attack and escape early in courtship and between mating and escape later on.

There remains a third, more controversial situation. The male and female of many species show no signs of aggression or fear towards one another at mating time. Sometimes there is no fighting at all (butterflies and whitefish); sometimes males fight other males but not females (fiddler crabs and swordtails). Many species in this category are without courtship. The others have been interpreted in various ways. Often observations suggest that the females are initially coy but responsive to stimulation. Such females usually prevent males from mating, without driving them away. An unwilling female fruit-fly, *Drosophila melanogaster* for example, fails to spread her genitalia or to part her wings; she even twists her abdomen away from the male, kicks at him, and flicks her wings. But the male stays close to such a female and continues to court until she responds. By contrast he turns away from other males or fertilized females when they perform similar, but more violent, repelling activities. Apparently the female simultaneously stimulates and prevents mating. Such a male is thwarted and he too might be expected to perform so-called conflict activities. Such a situation might be called M (thwarted). There are many reasons why females might be initially coy; perhaps the commonest is the need to prevent cross-mating which was discussed in the previous chapter.

Brown has demonstrated a situation of this type in the fruit-fly *Drosophila pseudoobscura* [47]. He analysed the situations in which a certain wing display (the V display) tends to occur. It is most common when mature males court immature (and unwilling) females. Similar males rarely show it when courting very receptive females for they mate almost immediately. Nor do immature males show it at all for they tend to stop courting whenever the female shows opposition. In other words the display is performed when a male with persistent sexual behaviour is thwarted in his mating attempts.

The discussion has so far centred upon males. Female displays also require explanation but here there is even less critical evidence. It is likely that the attacks and threats which a female receives from a male in early encounters induce both sexual and escape reactions in her. Evidence for sexual stimulation was discussed earlier (page 85) and females do flee if the attacks become too vigorous. Hence it is possible that the females of species which posture during early encounters (sticklebacks and jewel fish) are in an approach/avoidance conflict. Similar arguments can sometimes be applied to soliciting displays.

To sum up, there is evidence that courtship displays tend to occur when mating behaviour is simultaneously stimulated and opposed. There may not always be an exact balance between a mating reaction and an opposing one as in the idealized conflict situation, but it can be said that there is usually some interference with mating. Thus far the suggestion that courtship arose from behaviour induced by such interference is supported.

Conflict behaviour

Do courtship displays, in fact, resemble conflict behaviour? Conflict behaviour is defined by its occurrence in conflict situations and does not consist of unique activities; many of them occur in other situations too. However, many are distinctive and easily recognizable.

The most typical feature about conflict activities is that they are intermittent, incomplete and often distorted from their normal form. Very often an animal will alternate between the two opposing activities, performing each for only a short time and neither completely. Andrew illustrates this in his description of the behaviour of a hungry yellow-hammer perched near a food tray occupied by a dominant bird [1]. The food stimulates an approach for feeding, the dominant bird induces avoidance. The bird neither approaches nor retreats; in fact it alternates incomplete feeding movements with reactions typical of birds which are about to flee. The feeding consists of rubbing the mandibles as if breaking up food; 'fear' reactions consist of raising the crest, stretching the neck and flicking the tail (as if about to take off). The alternating behaviour is often called *ambivalent behaviour* [270, 26]; the incomplete activities *intention movements* [74]. Intention movements are also seen in other situations. For example, at the beginning of the nesting season a bird picks up sticks and drops them; this is incomplete nesting behaviour. In such a case, only the first stage of the activity is performed. But in conflict situations intermediate or late components of a sequence may occur without others. For a yellow-hammer, complete feeding behaviour would have consisted of flying to the dish, picking up food, rubbing it in the mandibles, and swallowing it. 'Fear' blocked all but the third of these activities.

Another typical conflict pattern is the *inhibited movement.* This usually occurs when a tendency to move away from a certain object or place is partially blocked by a tendency to stay there. A yellow-hammer, disturbed at the nest, may fly away very slowly and with a spread tail (a balancing device when flight is slow). Here the normal escape reaction is interfered with by the tendency of the brooding bird to stay on the nest. Sometimes this behaviour is elaborated into a distraction display in which the birds behave as if injured and draw predators away from the nest.

Certain positions or postures are also common in conflict situations. *Ambivalent postures* combine positions (of head, neck, limbs and feather or fur) typical of the two opposing activities. An oft-quoted example is the threat (and early courtship) display of the male herring gull which combines positions used in both attack and escape reactions. *Compromise postures* represent an intermediate position, where combination is impossible. In the bronze mannikin *Lonchura cucullata*, Morris describes how two evenly matched birds will stand parallel to one another in a flight, each twisting his head to spar with the other. A dominant bird faces his rival (to attack), a submissive one faces away (to escape). The compromise position appears to occur in birds in which attack and escape tendencies are evenly balanced [196].

Occasionally animals in conflict perform new and apparently irrelevant activities, although often incompletely and jerkily. A chaffinch *Fringilla coelebs* may peck viciously at buds in pauses during a fight, tearing them apart but not eating them; a herring gull, in a similar situation, may tug at nest material. Other examples are the fighting yellow-hammer which takes up the sleeping posture for a few seconds, the turkey *Meleagris gallopavo* which dips its beak into water but does not drink, and the zebra finch *Poephila guttata* which wipes its beak in the air but does not make contact with the branch. These activities have been called '*displacement*' *activities* [270] although the term is unfortunate because of physiological implications which are probably incorrect. Such activities have been thought remarkable because they are unexpected interruptions; unexpected, because animals tend to perform one activity at a time, even although others are stimulated. Parent birds, for example, may get very bedraggled while feeding young, yet they rarely stop to preen. However, it has recently been pointed out that sudden, incomplete bursts of a 'new' activity during the performance of another are not uncommon occurrences. It happens when one activity is 'taking over' from another and it happens when some especially strong stimulus for a new activity briefly occurs. It also

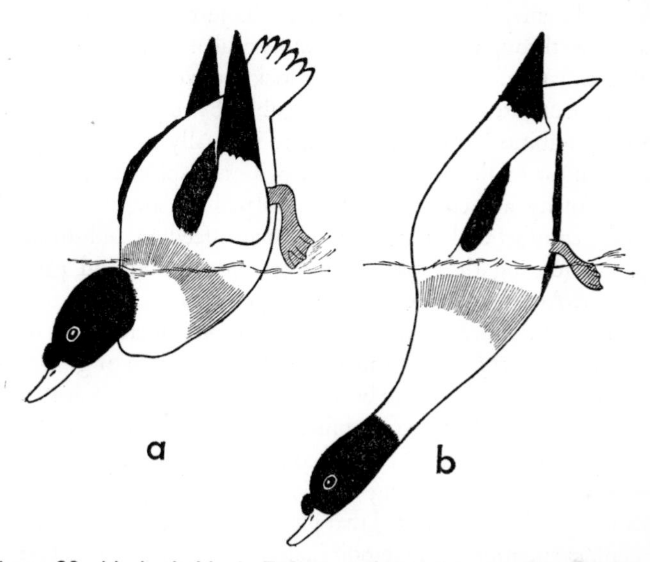

Figure 29. (*a*) A shelduck *Tadorna tadorna*, in a position commonly assumed during high intensity bathing (hindquarters lifted, forebody in water) and (*b*) in the feeding position which sometimes develops from it momentarily.
(After Lind [156].)

happens, as Lind has observed, when the first activity involves a posture or movement which is shared by the new one [156]. For example, a shelduck *Tadorna tadorna*, when bathing, often dips its head and neck into the water and lifts its hind quarters. This movement is also part of the feeding pattern, and occasionally bathing is interrupted by a quick feeding movement (figure 29). And nesting, by the blacktailed godwit *Limosa limosa*, involves pecking at the ground, again part of feeding. Frequently nesting is increasingly interrupted by feeding until the latter activity takes over. Lind argues that the common position itself provides stimulation for the new activity (perhaps through proprioceptive impulses from the muscles). And it may also bring the animal into close contact with stimuli releasing it (food or sand for digging). He argues that when this happens the new activity may temporarily 'break through'.

A 'displacement' activity, then, may be just one more example of a break-through occurring in the peculiar circumstance of a conflict situation. Perhaps the deadlock between two opposing activities facilitates this process (possibly by causing previously imposed inhibition to be raised). There is usually some stimulus to perform most of the activities described as 'displacements' in this context; many are continuously stimulated to some extent: for example, comfort and cleaning movements, feeding and drinking, sleep and (during the reproductive season) nesting and parental behaviour. And there is evidence to suggest that such activities can 'break through' more often if their stimulation is increased. Rowell has shown that artificial rain increases the occurrence of grooming by chaffinches *Fringilla coelebs* both in normal and conflict situations. Similarly bill-dirtying increases bill-wiping in both situations [240]. Comparable findings have been recounted for grooming by terns *Sterna* [127], and nest-fanning by sticklebacks [248] (figure 30). In addition, Morris has suggested that physiological changes may occur in a conflict situation (in metabolic activity, heart beat, blood circulation) which could stimulate feather and hair movements, comfort movements or drinking. An animal 'in conflict' may feel hot or cold, sticky with sweat or thirsty [1,194]. All this suggests that displacement activities are not unusual and extraordinary phenomena.

They are normal happenings occurring in special circumstances of which conflict situations are one example. In this chapter, however, we are not so much concerned with their physiology as with the fact of their occurrence during conflict and perhaps during courtship too.

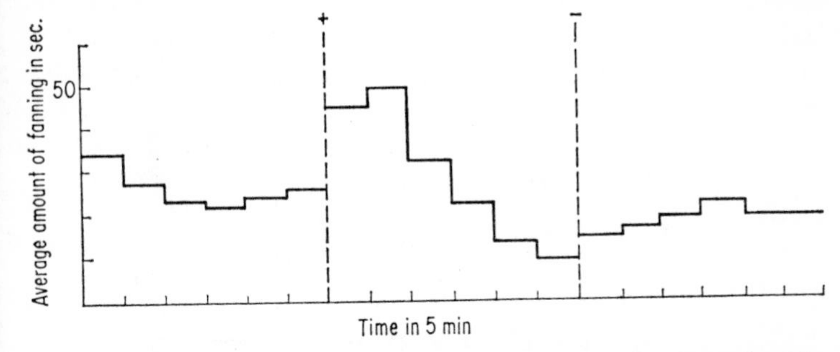

Figure 30. The effect of water containing concentrated carbon dioxide upon nest-fanning in the male three-spined stickleback *Gasterosteus aculeatus*, during the sexual phase. Fanning during this period is intermittent and directed at an empty nest: it is regarded as a displacement activity. During the parental phase the fanning ventilates the eggs in the nest and is stimulated by carbon dioxide from the eggs. + : introduction of carbon dioxide water; – : its removal. (After Sevenster [248].)

Finally there is *appeasement* or *submissive* behaviour [159, 160, 164, 275]. This occurs when a frightened animal cannot escape. Sometimes escape is physically blocked, sometimes it is 'opposed'; a frightened subordinate animal is still attracted to the group, and a frightened suitor to his mate. Most appeasement postures are as far removed as possible from aggressive ones. Weapons like teeth or beaks are turned away and aggressive display colours hidden. There is considerable observational evidence [102, 159, 245, 273, 275] (although few quantitative analyses) that such postures inhibit further attack, at least momentarily. Hence they are displays in their own right.

Resemblances between courtship and conflict behaviour are easy to find and are often striking. All zig-zag or pendulum movements or dances look very much like ambivalent behaviour. Examples are the zig-zag dance of the stickleback, the pivot dance of the zebra finch (figure 31) [191], the luring display of the guppy and the zig-zag approach of the male jumping spider towards his female. Tinbergen [208, 271] has interpreted the stickleback's dance in terms of quick alternations between attack and sexual (leading) behaviour. For 'zig' movements do sometimes end in bites and 'zags' normally resolve into excursions to the nest at the end of each dance phase. Inhibited movements are equally striking. They

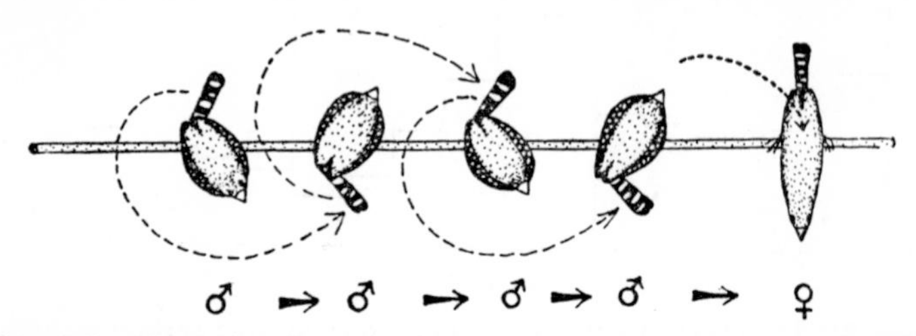

Figure 31. The pivot courtship dance of the male zebra finch *Poephila guttata*. The bird swings its body from side to side as it advances along a branch towards a female. Each swing turns the male away from the female, as if to flee from her, then brings him back towards her again. This suggests an alternation between fleeing and approaching. (From Morris [191].)

are seen in the sailing of the whitefish, the shimmering of the guppy and swordtail, the slow-motion flight of many female butterflies and the flight displays of the green heron. The lateral displays of partridges (figure 32), pheasants (Phasianidae) and fighting fish *Betta splendens* may well have developed in males which tended to adopt compromise postures before their females (a compromise between approach and retreat); circling behaviour can be similarly interpreted. Similarly appeasement postures are seen in the display of the female three-spined stickleback, the stretch display of the green heron, and the 'head-flagging' or 'facing away' of black-headed gulls [275]. This movement – turning away of the aggressive black cap and beak – is used as a greeting between mates (figure 33). Finally displacement activities are common both as interruptions to courtship and as parts of displays. Examples of interruptions are seen in the prey-shaking movements made by chars *Salmo alpinus*

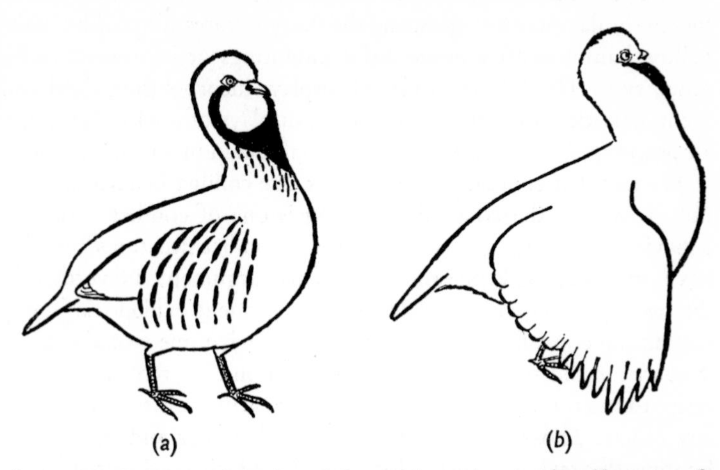

Figure 32. The lateral display of the red-legged partridge *Alectoris rufa*. In this posture the male walks past or around the female keeping always sideways on to her and thus displaying his face mask and barred flank markings. The wing away from the female is trailed and spread, perhaps for balancing.
(*a*) Typical display. (*b*) Intense display seen from the opposite side.
(From Goodwin [95].)

Figure 33. Mutual 'greeting' displays between mates of the black-headed gull, *Larus ridibundus*. Initial threat (*a*) is followed by head-flagging (*b*), turning away the mask and beak.
(From Tinbergen [273].)

when females escape them [84], the cleaning activities of male fruit-flies in similar circumstances, and the sleep postures adopted by male yellow-hammers after unsuccessful mountings or incomplete bill-raised runs. The best examples of displays incorporating displacement activities come from the pairing displays of ducks. These are all preening and bathing movements elaborated in various ways.

The fact that courtship displays look like conflict behaviour does not prove that the courtship situation is one of conflict. Conflict behaviour is defined by its occurrence in conflict situations and can have other causes; hence this is a circular argument. Nevertheless the resemblance gives weight to the independent evidence of conflict during courtship already discussed. It is what one would expect if courtship did originate in conflict situations. Moreover it seems easier to interpret the form of many displays in these terms than in any others. Those displays of the yellow-hammer and the snow-bunting *Plectrophenax nivalis*, for example, which are directed away from the female, make their best sense if regarded as the outcome of an approach/avoidance conflict in which avoidance is temporarily dominant.

Effective displays

We must now ask how conflict behaviour might evolve into a courtship display. Displays are behaviour patterns which influence the behaviour of other animals. Any activity may sometimes do this; for example, an animal feeding may induce others to feed; but a display must normally and regularly do so. And the influence of a courtship display must be such as to improve mating.

The initial influence of some of the conflict activities described above can sometimes be guessed. Many of them are conspicuous; for example, ambivalent movements by virtue of their rhythmical alternation, and inhibited movements because slow-motion involves exaggerated body undulations or wing-beats and the spreading of tail feathers or fins for braking. And these, together with all unusual, jerky, hesitant and incomplete activities tend to attract attention. This in itself is important in mating. Lorenz has also suggested why intention movements (which are seen everywhere in courtship) may influence other animals [159, 162, 163]. Many higher animals (particularly social ones) quickly learn the postures and movements associated with the normal activities of their group. Hence they recognize intention movements for what they are and react accordingly. Intention movements of mating and sometimes nesting thus become sexually stimulating; appeasement postures or escape reactions diminish attack or escape tendencies which may be interfering with mating. Even aggression may become sexually stimulating because of its association with mating behaviour.

But if a conflict activity is to become an effective display, its stimulating and attention-catching features must be enhanced. Comparison of displays with their supposed origins suggests how this may have happened. Such comparisons are particularly convincing in closely related groups of animals, where a given activity is modified in different ways in different species. Lorenz shows how different features of preening behaviour have been exaggerated in different species of duck (Anatinæ) in the evolution of the mock-preening display [162, 163]. The display of the mallard *Anas platyrhynchos* closely resembles normal preening: the beak is drawn along the underside of the partly lifted wing (figure 9(*b*), page 41) producing a loud Rrr sound; there is also a colour effect for a blue speculum is revealed when the wing is lifted. The mandarin *Aix galericulata* enhances this effect by exaggerating the wing movement, raising it like a sail at each outburst of display to show off the large rust-red tertiaries (figure 34); at the same time it reduces its beak movement to the touching of a bright orange secondary. By contrast, the shelduck *Tadorna tadorna* has elaborated the sound effects: its beak movement has become a powerful rapid stroke along the shafts of the wing quills producing a low rumbling sound.

This process is one of selective *exaggeration*; another common modification, illustrated by this example, is that of *simplification*. While effective elements are exaggerated, unimportant ones may be simplified or even omitted. The mandarin has reduced the beak movement in its preening; so has the wood-duck *Lampronessa sponsa*, which merely touches one conspicuous feather, with an extremely rapid jerky movement quite unlike preening.

Figure 34. Drinking followed by mock-preening in the mandarin *Aix galericulata*. Very often, when mates meet, both drink; then the drake mock-preens, touching the wing feathers on the side nearest the duck. The hood plumes are raised for drinking and stay up for mock-preening.
(After Lorenz [162].)

Changes in speed or relative timing are often involved in exaggerations. Movements are speeded up when they thereby produce startle effects, or are slowed down to prolong particularly effective postures or colour displays. In the preliminary shaking display of the mallard, both timing and co-ordination are changed. It will be remembered that the presumed original feather-shaking activity usually consists of two parts, 'high swimming' and 'thrusting', which follow closely upon one another. In display, the conspicuous high swimming phase is prolonged and the thrust is repeated three times thus adding rhythmical movement to its attention-catching qualities. Very different speeds of leg waving are also seen in the various species of fiddler crab, some producing flash effects from the brightly coloured claws, others varying the tempo and freezing the claw briefly in its fully extended position.

Rhythmic repetition is a common and predictable feature of displays, for most sense organs respond more readily to spaced, repeated stimuli than to steady, static ones. Rhythmic movements may produce visual sequences (fiddler crabs and sticklebacks), tactile effects (undulating fish), or sound patterns (grasshoppers, woodpeckers *Dryobates* and domestic cocks (with their wing-clapping)).

The association of activities with similar effects may increase the influence of each. This is particularly liable to happen where an already existing display occurs in the courtship situation. Tinbergen suggests that the upright posture (with vertical, stretched neck) has become linked with head-flagging in the courtship of the black-headed gull. The upright is used in fights by birds who are more likely to escape than attack; head-flagging is an appeasement posture; hence they may augment one another in diminishing attack reactions in the partner [273]. Similarly, drinking is associated with mock-preening in the greeting ceremonies of many species of ducks [162, 163]. Drinking seems to be a friendly greeting signal used by many ducks on all occasions. Mallards only occasionally associate it with mock-preening in courtship, but other ducks always do so: sometimes drinking precedes mock-preening (wood-duck and mandarin) (figure 34) and sometimes it follows it (gadwall *Chaulelasmus strepara.*).

The association of displays with coloured or conspicuous anatomical features is another expected and well-supported development. The colour patches of birds, in particular, are frequently puffed up and outlined by contrasting borders; crests and wings are raised, bodies are twisted to display stripes or patches; tails or wings are flicked to flash under-markings.

This is particularly striking when, in closely related species, different modifications of the same original display activity are associated with special body features. Among the jumping spiders, the legs which are thickened or coloured are moved more frequently and more conspicuously than the others; these vary between species. Similarly, while most gulls occasionally open their mouths during threat displays, the kittiwake *Rissa tridactyla* always does so in an exaggerated way; it has a bright orange lining to its mouth [273]. And the hooded gull exaggerates the use of a forward threat posture (found occasionally in other species), in accordance with the possession of a dark facial mask [273]. It is not very profitable to argue whether colour features preceded display modifications or *vice versa*. Very often they must have evolved together.

It is also important that a display should be regular and stable in its occurrence if it is to give a clear and unambiguous signal. It must become a normal, instead of perhaps an occasional, preliminary to mating. In addition, Morris [197] has stressed the importance of stability in what he calls the development of *typical intensity*. Many behaviour patterns vary widely in intensity and completeness. Early in the season, a nesting bird mainly picks up twigs, later it also carries them to the nest-site and builds them into a nest. Moreover both carrying and building may vary in completeness. Displays are much more stable: they tend always to have the same form, speed and rhythm; they are usually performed completely or not at all. There seems little doubt that this is important for their easy recognition.

Interactions

In the above paragraphs, display has been treated as an isolated characteristic, evolving *in vacuo*. But it is the whole animal, not each separate character which evolves, and the consequence of any change must be considered in relation to every other feature of its life.

The need for animals to conceal themselves must often interfere with the evolution of displays. Displays are most effective when they are most conspicuous, but a gaudy animal, even if irresistible to females, will not leave many offspring if he falls an easy prey to predators. Hence very brightly coloured animals have few predators or good places of concealment. Sometimes bright display colours are adopted only at the breeding season; or they may be concealed by feathers or fur, to be displayed only in special postures (figure 35). A striking example is the great bustard *Otis tarda*, in which a sandy-coloured male is suddenly transformed into a snowy white one by the turning over of wing coverts and tail feathers [289].

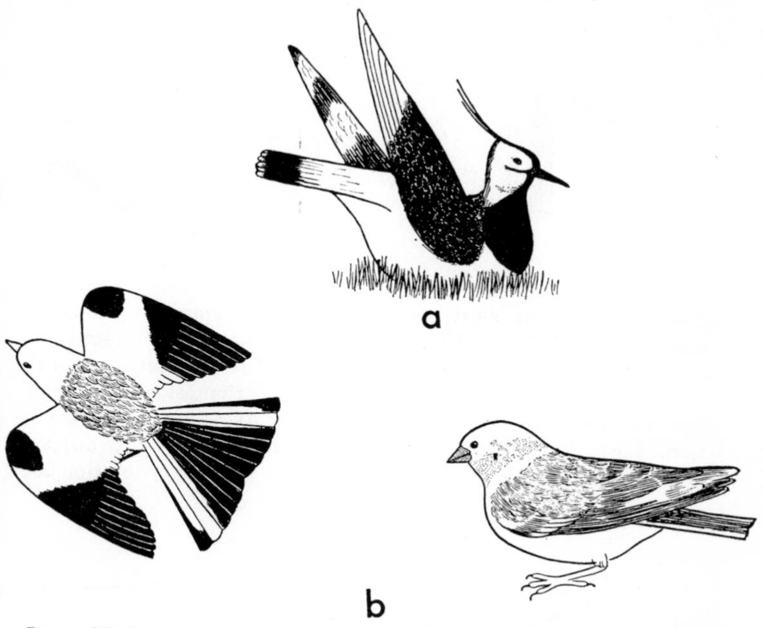

Figure 35. Special postures of courting birds revealing striking markings which are otherwise mainly concealed.
(*a*) The display of the undertail coverts of the lapwing *Vanellus vanellus*; (*b*) the display of tail and wing markings of the snow bunting *Plectrophenax nivalis subnivalis*.
((*a*) After Tinbergen [269]; (*b*) After Tinbergen [267].)

The general way of life of a species may also influence the form of its courtship. Of two closely related species of fruit-fly, *Drosophila melanogaster* and *D. simulans*, the former is much more active (for reasons unknown). Hence *melanogaster* males usually have to court their females on the run, for the latter rarely stay still when approached. Wing-vibration is therefore the most effective feature of the display, for it enables the male to waft air vibrations towards the female even while he is pursuing her. By contrast *simulans* males usually court a stationary female: they can circle around her and give a visual display in front of her. Hence *melanogaster* stresses vibrational elements, *simulans* visual elements, in a basically similar display [171] (figure 36).

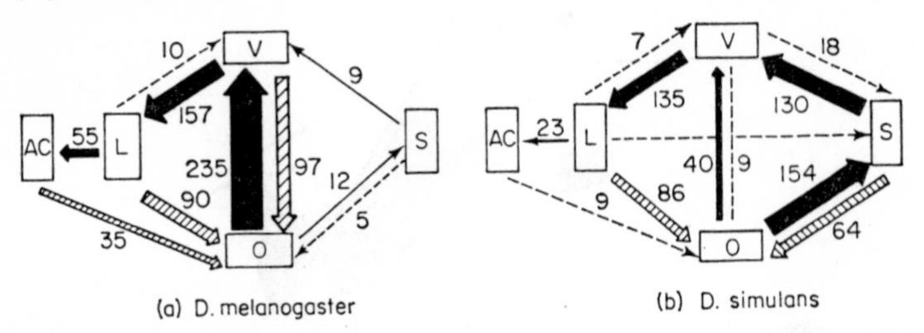

Figure 36. Diagrams illustrating courtship sequences in *Drosophila melanogaster* and *D. simulans*. The arrows indicate the direction of sequences and their thickness is proportional to the frequency with which each sequence was observed. O = orientation; V = wing vibration; S = wing scissoring; L = licking; AC = attempted copulation. Scissoring (which frequently develops into a visual display) commonly occurs in *D. simulans* between orientation and vibration but it is usually omitted in *D. melanogaster*.
(After Manning [171].)

A second example comes from Cullen's studies of the kittiwake *Rissa tridactyla* [72]. Unlike most gulls, this bird nests on the narrow ledges of steep cliffs. Good sites of this type may be rare, and they tend to be claimed early in the season and defended closely. All pairing, courtship and mating occurs on this narrow ledge, close to the nest. Correspondingly, a particular display, called choking, which occurs always in relation to the nest, is unusually common in kittiwake displays (figure 37). It appears as an advertisement in unpaired males, as part of the greeting ceremony at pairing, and (in modified form) as an appeasement display at the end of greeting (at the point where the black-headed gull head-flags). Most gulls pair away from their future nest-sites; they rarely perform choking at pairing time and have quite different displays for advertisement and greeting.

The evolution of displays must also be influenced by sensory capacities. Sometimes, particularly among the invertebrates, related species (or families) tend to employ different sense organs. For example, Crane finds a trend among jumping spiders from hunting by smell to hunting by sight. This trend is paralleled in courtship with an increase in visual displays (leg and body movements and colour effects) [67]. However, not all differences in the type of stimuli employed in courtship reflect sensory changes. They may depend on differences in habitat (some court in darker places than others), in form (some have special structures to display) or in habit (such as the effect of activity in fruit-flies).

Figure 37. 'Choking' in the kittiwake *Rissa tridactyla*. A display always performed near the nest. (From Tinbergen [273].)

Courtship may also be influenced by other displays used by the same animal. It has already been suggested that appeasement displays are sometimes incorporated in courtship because they have effects which are useful in the mating situation. By contrast, courtship must be as distinct as possible from displays with unwanted effects. In particular, striking colours and movements used in aggressive displays must be avoided where it is important for courtship displays not to evoke agonistic behaviour. The kittiwake, which uses its bright orange mouth as an aggressive signal, always closes it in 'friendly' displays [273]. And the green heron, inviting the female to its nest, makes soft calls in contrast to its fierce aggressive cries.

Confusion must also be avoided with the displays of other species, especially closely related ones living in the same area. The highly distinctive displays of many species of fiddler crab all on the same beach illustrate this. Often the displays or calls of sympatric species diverge while others do not. For example, populations of two species of frog, *Microhyla olivacea* and *M. carolinensis*, have distinct calls in regions where they overlap but identical ones elsewhere [35, 36, 37]. There are greater differences between the songs of overlapping species of *Hylocichla* thrushes than of others [254], and female grasshoppers discriminate against the songs of overlapping species but not against others [209].

Ritualization and emancipation

The evolutionary changes outlined above often modify a display considerably from its original form. Julian Huxley [124] has called these changes *ritualization* – an effective term which gives some impression of the formalized, 'unnatural', end-result.

A feature of ritualization not so far discussed is that called *emancipation* [38, 270]. This term has been given to the hypothetical process whereby during evolution a behaviour pattern becomes 'freed' from its original set of controlling factors and is transferred to another. For example, duck courtship-preening, if fully 'emancipated', would be an integral part of the mating system, influenced solely by sex hormones, female stimuli and other sexually significant factors. It would not be affected by feather wetting or feather disturbance. Where conflict movements or postures are involved, as in the alternately aggressive and sexual phases of the stickleback's zig-zag dance, then a fully emancipated display would cease to depend upon the conflict state and would vary with sexual state, irrespective of aggression.

The idea of emancipation at first seems highly plausible. Certainly courtship vigour and persistence in courtship varies with sexual responsiveness in general; with the frequency of sexual approaches and mating attempts; with the speed and ease of eliciting such responses. In seasonal animals, courtship and sexual behaviour tend to increase together at the beginning of each breeding season and fall together at the end of it. In the stickleback, courtship frequency (the number of zig-zags per minute) increases with leading. The latter is the movement whereby a male guides a female to its nest and is treated as a purely sexual activity (part of the mating process). Hence, since they vary together, it is reasonable to suppose that courtship and mating are controlled by the same factors.

In the three-spined stickleback analysis has been taken further and suggests that zig-zags are 'emancipated' from a conflict situation [248]. Early in the reproductive cycle (during nest-building), a female, presented in a jar, is attacked: there are no zig-zags. Later there are a few zig-zags, a few leadings and a few bites. Later still there are many zig-zags, many leadings and virtually no bites. Hence over a period when aggressive reactions are decreasing and sexual ones increasing (towards the female) the frequency of courtship responses closely parallels that of sexual ones. This is not the result to be expected if courtship depends on conflict. The point of balance between aggressive and sexual reactions (which in this case are opposing ones) should occur in the intermediate period. Hence courtship should be most frequent in this period; later, one would expect relatively few zig-zags and many leadings.

However, this argument is not entirely watertight. In the last phase, given a willing female, a stickleback male is likely to proceed immediately to mating after a few brief zig-zags. In the experimental situation (and sometimes the natural one too) he is thwarted in this by the failure of the female to follow. Hence, as in the fruit-fly (see page 102, courtship in this phase might be related to the thwarting of mating reactions, and thwarting is likely to increase with the strength and frequency of mating attempts.

Moreover, the courtship of some animals at least is more related to interference with sexual activities than to sexual activities alone. There is Brown's analysis of the fruit-fly V display mentioned above, suggesting that both high motivation and thwarting are necessary for its occurrence. Other evidence has also been discussed earlier: the pre-copulatory displays of many birds (such as the yellow-hammer and house-sparrow) are sporadic and may be omitted: in some species, such as the green heron, they die out with experience: and domestic cockerels in their home pens, among familiar hens, display less frequently than cocks in strange pens (although they attempt to mate just as often) [292]. In all these cases omissions or abbreviations of courtship are associated with situations where sexual responsiveness is high in both partners and opposing factors low. In other words, courtship occurs only in situations where interfering factors do exist.

We cannot, however, dismiss the idea of emancipation. There is no reason why it should not have occurred. Moreover it is not necessary to suppose any major switching of control systems. The activity in question might become responsive to sex hormones; this would be one way of assuring that it occurred habitually (instead of occasionally) in the mating situation. In this sense, then, it is perhaps misleading to call the process emancipation, and since it is, in any case, exceedingly difficult to demonstrate, it remains a concept which should be treated with caution.

7: Genetics

If courtship is to evolve, it must vary. The variation must affect reproductive success and it must be inherited. So far we have discussed the advantages of certain variations. What do we know of their determination and their inheritance?

Every characteristic of an organism is the result of an interaction between inherited genetical potentialities and the environment. It is nonsense to ask whether any given feature is determined by genes or by the environment. It is always determined by both, just as the nature of a cake depends both upon its ingredients and the cook's treatment of them. It is however perfectly sensible to ask whether *differences* are genetically or environmentally determined, although it may be difficult to provide an answer. A cake may be modified by changing the ingredients or the cooking but it is not always easy to determine which has occurred by examining the cake.

Their problem of distinguishing between genetical and environmental effects has been discussed by Haldane [101]. It varies with the characters concerned. Some, such as eye colour and coat colour, are relatively unaffected by the environmental changes which normally occur. The variation is mostly genetically determined. But many characters, like stature, vigour, activity, resistance to disease and particularly behaviour, can be modified by circumstances like food availability, contact with other animals and experience which may vary considerably from individual to individual. Hence, two populations of insects may each vary in body size, for instance, but for different reasons. One may consist of genetically similar individuals which compete for food; the other may be genetically variable and have plenty of food. In the first population the variation will be mainly environmental in origin; in the latter, it will be genetical.

Usually, of course, both factors operate. The situation is further complicated by the fact that some genotypes (particularly very homogeneous ones such as are found in inbred lines) are more affected by environmental changes than are others and some genetical differences express themselves more in some environments than in others.

When different species are compared, the difficulties are even greater. There may be consistent environmental (as well as genetical) differences between two populations. Hence stable differences between two species are not necessarily genetical as is often assumed. Bird songs, for example, are highly specific and usually stable, yet not all their characteristic differences are gene-determined. Birds of some species sing atypically if they do not hear their 'own' songs in early life and, if they hear other songs instead, they modify their performance accordingly (although not completely) [265]. How can all this be sorted out?

Problems of method

We will consider methods which have been used to study genes and behaviour in general, before applying them to courtship in particular.

Genetical influences upon behaviour are most easily studied within interbreeding populations. A direct, but rather hit-or-miss, method is to examine animals possessing some known mutant gene (usually with an easily detected effect) to see if the unusual gene changes behaviour. This is done by comparing mutant animals with normal ones (otherwise similar genetically) under conditions which are as nearly identical as possible. Alternatively, it can be arranged (with large numbers of animals of each type) that uncontrollable environmental factors are shared equally between the two groups. Consistent behavioural differences between the two groups are then likely to depend upon the changed gene.

This method is based on the knowledge that one gene can affect many characters and that many genes influence behaviour. Where it has been used, it has often proved successful, as has the rather similar method of comparing inbred or selected lines of animals for incidental behavioural differences. Inbred lines, usually obtained by repeated brother–sister matings, are as a rule very uniform genetically. Selected lines are populations artificially selected for some particular trait, such as body size or coat colour. They differ consistently from unselected lines, or from lines selected for an opposite trait (small *versus* large, or brown *versus* black), at least by the genes determining these traits. Similarly two distinct inbred lines will differ consistently according to the genetical differences between the stocks from which they have been derived. And consistent behavioural differences between two selected or two inbred lines (developed and tested under parallel conditions) probably reflect these genetical differences. It is often possible to prove this by crossing the lines and examining the hybrids, for genetically determined differences will segregate among the offspring in a predictable way.

It is also possible to select directly for a behavioural variant by choosing, as parents for the next generation, the animals which show it most strikingly. One might choose, for instance, the most active animals. If the difference is mainly genetically determined, then the offspring of such parents will be, on the average, more active than those of unselected parents. The rate at which average activity changes in each generation can, in fact, be taken as a measure of the genetical component of the variance. If most of the differences are gene-determined, selection will be rapid, but if many of them depend upon environment, it will be slow and oscillating. This method also tells something of the number of genes affecting the character for the pattern of selection varies accordingly. Again genetical differences between selected and control (not selected) populations can be studied by crossing the two. For a fuller discussion of this method see Falconer's book [86].

Selective breeding is, however, often impossible, for example, in man or in other slow-breeding species. Here the genetical basis of variation is often assessed by studying relationships. Related animals (members of one family) are likely to have more genes in common than are unrelated ones. Hence genetical differences are likely to be less pronounced within family groups than between members of different families. Similar comparisons can be made between twins as opposed to other siblings or, better still, between uniovular, as opposed to binovular, twins. This applies especially to man but also to other mammals, such as the Texan nine-banded armadillo *Tatusia novem-cincta*, which always has uniovular quads. Uniovular twins are genetically identical; binovular, like ordinary siblings, share roughly half their genes in common. Twin studies are, in fact, preferable, because the environment of each pair of twins is likely to be very similar indeed. In family studies it is clear that environmental as well as genetic factors are likely to be more similar within family groups than outside them; hence the case for genetic influence can only be made here where environmental differences between families can be assumed to be small or of negligible effect.

The nature of differences between species and races can only be demonstrated satisfactorily when it is possible to interbreed and study the pattern of inheritance in the offspring (as described for inbred and selected lines; again see Falconer). A study of relationships may suggest which are genetical traits but it is often difficult to assess the influence of environmental factors. Ideally, hybrids should be bred to the second generation or they should be backcrossed to the parent populations; this is only occasionally possible between very closely related species. Frequently hybrids, for various reasons, will neither court nor breed. Nevertheless a few valuable studies of courtship inheritance have been made by this method.

Variation between species

Where courtship has already evolved we cannot study the nature of the process which brought it about. But there are three lines of enquiry which can suggest the genetical changes which may have been involved:

1. What kinds of differences exist between very closely related species, and are they genetically determined?
2. Do genetically determined differences of a similar nature exist within species?
3. Are there, in any species, genetically determined differences in conflict or other behaviour which could account for the origins of courtship?

Most of the work so far, has concentrated on the first of these questions. Differences between the displays of related species which are likely to be of significance in courtship evolution have already been discussed. They include variation in form, timing, patterning, elicitability and regularity of occurrence. We are here concerned with very closely related species (as opposed to more distantly related ones). In these we see the first stages of courtship divergence and we should expect to find the variations concerned within, as well as between, species.

One conclusion to be drawn from these comparisons is that large differences in the *form* of specific display activities are rare at early stages; but differences in the frequency of their occurrence and in their rhythm and timing are very common indeed. For example, among the fiddler crabs, the group of vertical wavers appears to be distinct from the American group of lateral wavers. Hence the difference in the form of the wave exists between two more distantly related groups: within each group the species differ from one another mainly in the speed and relative timing of the various parts of the display [68]. Comparisons between other species groups tell the same tale. Considerable differences in the type of movements employed, exist between the fruit-fly subgenera *Drosophila* and *Sophophora*; but within each group, and particularly within groups of closely related families, the movements remain the same but differ in the relative frequency of their occurrence as well as in timing of the whole display [253]. This is well illustrated by the comparison between the displays of *Drosophila simulans* and *D. melanogaster* illustrated in figure 36, page 116. Among birds, too, comparisons

Figure 38. The forward threat display (a) of the black-headed gull *Larus ridibundus* and (*b*) of Hartlaub's gull *Larus novae-hollandiae hartlaubi*. ((*a*) From Tinbergen [273]; (*b*) From Tinbergen and Broekhuysen [274].)

of related species of gulls [273], finches [113] and ducks [162] all illustrate the prevalence of differences in frequency and timing.

Where form differences do exist they are relatively small and are of two main types. There may be differences in intensity or completeness; legs may give a bigger swing, feathers or fins be extended more completely. Alternatively where there are ambivalent postures or alternating intention movements, one species may have a more marked expression of one reaction than another. One of the best analysed examples is Tinbergen's comparison of the courtship of Hartlaub's gull *Larus novae-hollandiae hartlaubi* with that of the black-headed gull *Larus ridibundus* [274]. In almost every display Hartlaub's gull exaggerates the fear reactions. In head-flagging, there is a thinner neck and sleeker plumage; in the 'forward posture' there is an upward-pointing bill (figure 38). These are all features which normally accompany escape reactions. Moreover both alarm calls and the displays associated with escape are commoner in this gull than in the black-headed gull.

There are now a number of investigations demonstrating a genetical basis for variations such as these. Hinde studied three species of finch and some of their hybrids [114]. The canary *Serinus*, the greenfinch *Chloris chloris* and the goldfinch *Carduelis carduelis* have very similar displays, differing mainly in the frequency of display elements which are common to all. For example, the canary has 'wings-raised' courtship displays which may vary from an 'aggressive' form to a 'non-aggressive' variety (figure 39). The greenfinch has similar displays, but here the non-aggressive forms are more common (figure 40). In the goldfinch, none of these displays are common although they do occur. Nest-calling displays are frequent in the goldfinch, less common in the greenfinch and rare in the canary.

Form differences resemble those found between the gulls. The greenfinch and the goldfinch show more fear reactions than the canary. There is a more marked raising of the head and more fluffing of the neck feathers in the greenfinch's 'sleeked wings-raised' display: there is also greater fluffing of the feathers during the period of copulation attempts (figure 41). Fluffing is even more common in the goldfinch (figure 42). By contrast, many of the canary's displays

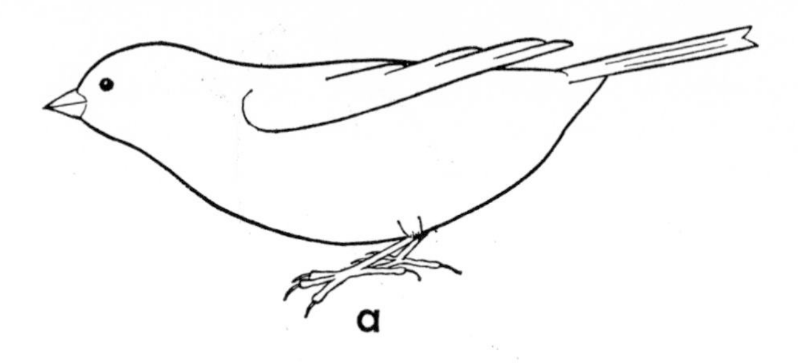

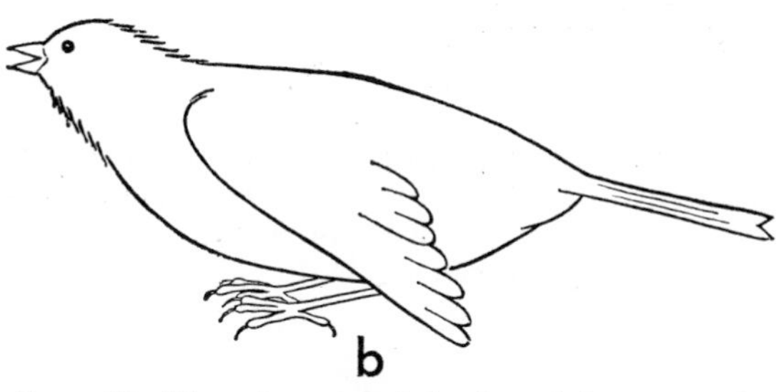

Figure 39. The wings-raised display of the canary *Serinus*.

(*a*) Aggressive form with sleeked feathers and horizontal body; (*b*) non-aggressive form with raised, fluffed head and neck. In posture (*a*) a male faces his mate and may attack her; in (*b*) he stands laterally and hops around her, sometimes carrying nest material.

(After Hinde [113].)

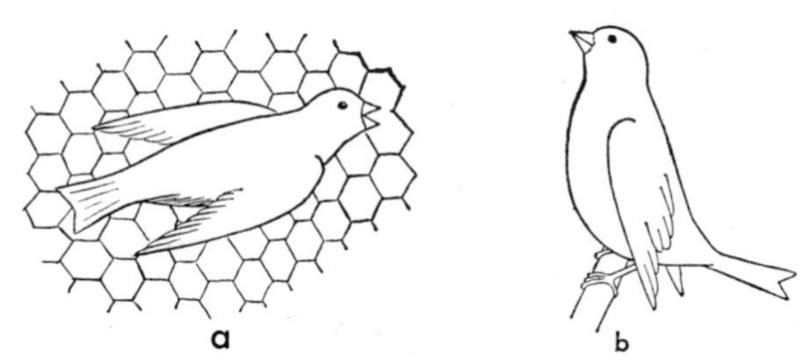

Figure 40. Aggressive and non-aggressive forms of the wings-raised posture of the greenfinch *Chloris chloris*.

The aggressive form (a) is common only early in the season: it is superseded by the non-aggressive form (*b*) which is more upright and more fluffed than that of the canary.

(After Hinde [113].)

Figure 41. The fluffed posture of the greenfinch, commonly assumed by the male when he is subordinate to his mate. He sits in this posture between attempted mountings.
(After Hinde [113].)

Figure 42. A fluffed chase in the goldfinch *Carduelis carduelis*. Fluffing is much more common in this species: it occurs on the same occasions as in the greenfinch but on additional ones too, as when mates chase one another.
(After Hinde [113].)

emphasize typical aggressive features like sleeked feathers and horizontal body.

Hinde found hybrids to be intermediate between the parent species with respect to most of these differences (qualitative and quantitative). Occasionally hybrids approach (but are never identical with) the condition in one parent rather than another. For example, nest-calling displays are common in all goldfinch hybrids but not so common as in the goldfinch itself. On the other hand, whenever displays are identical between species, they appear unchanged in the hybrids. In Hinde's experiments, the birds were kept in very similar conditions, and hybrids were reared by their parents (of two different species). Hence the evidence strongly suggests a genetical basis for the differences, although environmental, particularly parental, influences cannot be ruled out. The intermediacy of the hybrids also suggests, although it does not prove, that many genes are involved. Unfortunately, no further crosses were made to establish this.

One feature of particular interest in Hinde's work concerns 'pivoting' in the goldfinch (figure 43). This is a stereotyped display: the male swings his body from side to side in a regular rhythm. The

Figure 43. The pivot dance of the goldfinch. The regular swinging of the body is first towards and then away from the female. It suggests alternating intention movements of attacking and fleeing from the female.
(After Hinde [113].)

other two species have no such display, although both make occasional, irregular and incomplete swings towards or away from the female. Goldfinch hybrids (but not others) also swing quite a lot, but again the movements are less marked and less regular than in the goldfinch itself. This suggests that genetical changes in the goldfinch have developed pivoting as a display from the rare intention movements typical of the other two species.

A similar type of investigation by Dilger on African parrots has given very similar results [80]. Most of the differences between the displays of *Agapornis roseicolis* and *A. fischeri* are differences in frequencies of the various display elements. Again the hybrids show intermediate frequencies and sometimes resemble one of the parents. As an example, a display called 'switch sidling' accounts for thirty-two per cent of the total courtship in *roseicolis*, fifty-one per cent in *fischeri* and forty per cent in the hybrids.

Hormann-Heck describes a qualitative difference between crickets, which has resulted from a frequency change [119]. *Gryllus bimaculatus* makes pre-courtship sounds by raising its elytra repeatedly; *G. campestris* makes this movement only once so that there is no resultant sound. Most of the hybrids have intermediate behaviour, producing a sound but only a single one. From the second generation hybrids and some backcrosses, Hormann-Heck concluded that this behavioural difference, unlike most of those so far quoted, is the result of a single gene substitution between the species; the intermediate hybrids are then heterozygous.

A final example comes from studies by Clark, Aronson and Gordon on the inheritance of a form difference in the displays of the platyfish *Xiphophorus maculatus* and the swordtail *Xiphophorus helleri* [60]. It concerns the nature of the movements made after the female has been approached. The platyfish simply backs away with body limp and fins folded. The authors call this *retiring* and contrast it with the swordtail's more specialized shimmer-swimming across the female's head which they call *backing*. The inheritance patterns are as illustrated in figure 44. The results suggest that the differences depend upon many genes, with some dominance of those concerned in the more specialized swordtail pattern.

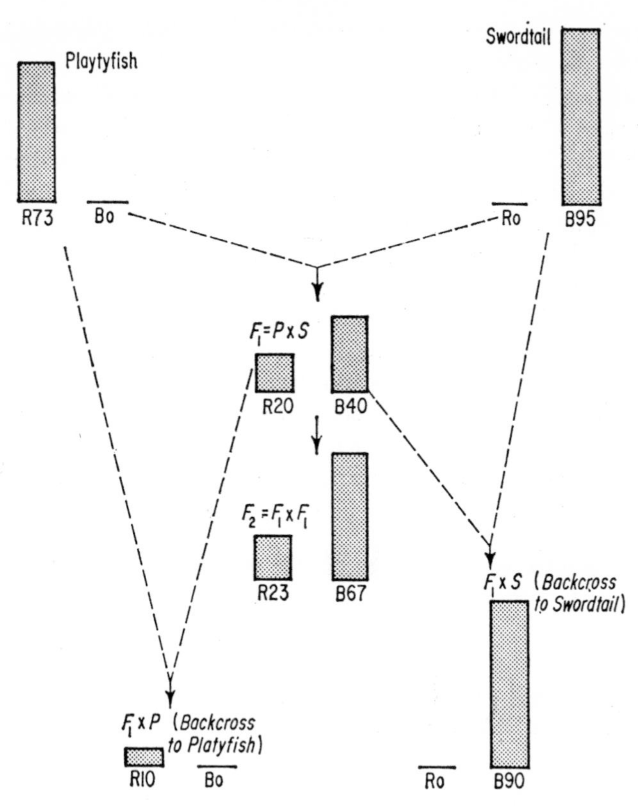

Figure 44. Diagram illustrating the pattern of inheritance of two different forms of a courtship movement: *retiring* in the platyfish *Xiphophorus maculatus* (R), and *backing* in the swordtail *Xiphophorus helleri* (B). The columns and figures indicate the percentage of males showing either movement in F_1, F_2 hybrids and backcrosses. (From data given by Clark, Aronson and Gordon [60].)

Variation within species

These investigations suggest that much of the divergence of displays between species is due to the natural selection of genetically determined differences. If so, we may expect evidence of similar genetically determined variation between individuals within each species. Departures from the typical will usually be disadvantageous, since the typical is a product of natural selection. If, for example, two species differ in the frequency of performance of a certain activity and it seems as if a given frequency has been selected as the most effective in each case, we should still expect to find within each species a range of genetically determined variation above and below the optimum, for the character must be genetically variable, otherwise selection could not have occurred.

Most observations suggest that there is considerable individual variation in courtship, in spite of its stereotypy, but few attempts have been made to discover how much of this is genetical. In the fruit-fly *Drosophila melanogaster* however, studies of the effects of single gene mutations have been made. Insects with a single mutant gene, yellow (changing the body colour), tend to have longer bouts of orientation and shorter bouts of wing-vibration in their courtship than have normal flies (figure 45), although the form of both

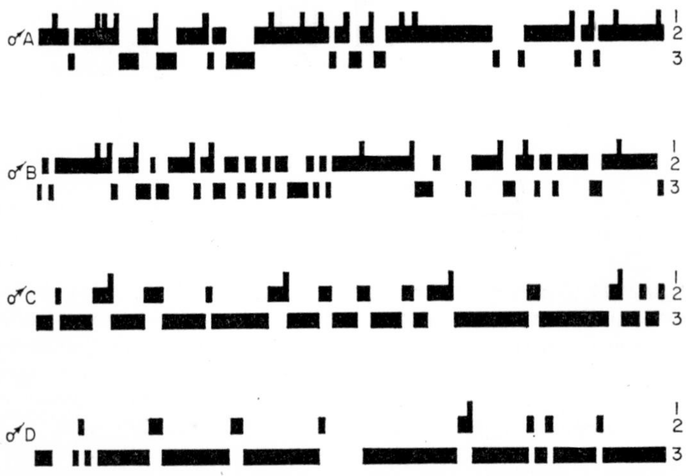

Figure 45. Different patterns of courtship shown by males of the fruit-fly *Drosophila melanogaster*. Each diagram represents a sequence of events in time. 1 = licking; 2 = vibration; 3 = orientation. The sequences A and B with long bouts of vibration (and licking) are typical of normal flies; C and D with shorter vibration bouts and longer orientation bouts are typical of yellow flies. (After Bastock [24].)

these activities is the same [24]. This makes yellow males less effective courters, for wing-vibration is the most stimulating element in the display of these insects [24]. Another body-colour mutant, 'ebony', affects the continuity of courtship: ebony males tend to lose their females easily while courting; they allow them to escape and courtship is interrupted while they search for them [71].

Ewing compared two populations of this insect which had been selected for large and small body size respectively [83]. He found that the displays of the large insects contained more orientation and less vibration than normal, while those of the small insects showed the reverse effect (figure 46). Since the populations had been kept

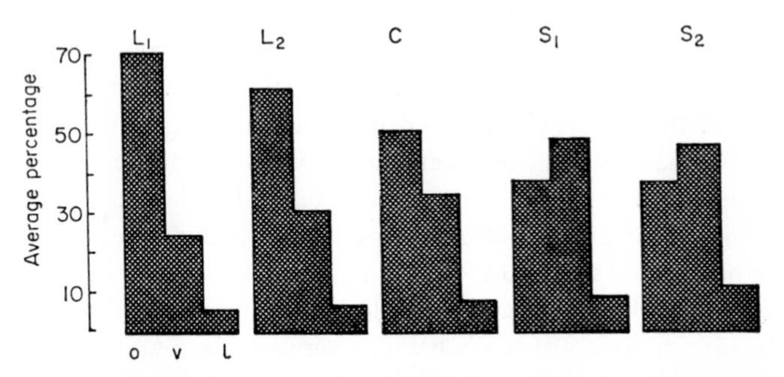

Figure 46. Histograms showing the average proportions of the three main courtship activities shown by males of five different populations of *Drosophila melanogaster*. L_1, L_2, selected for large body size; C, control population; S_1, S_2, selected for small body size. o = orientation; v = vibration; l = licking.
(After Ewing [83].)

under strictly similar conditions (and each was duplicated) these courtship differences were presumably a result of the genetical changes resulting from selection for size. In all these cases the genetically determined variation concerns proportions and patterning of separate elements in a display. This is the type of variation most commonly found between species; hence these findings, although limited, agree with expectation.

The origins of courtship

Most of the heritable differences so far discussed suggest how pre-mating behaviour, once established, can become modified into an effective display. But what of the origins of that pre-mating behaviour? For a favourable activity to become a usual, instead of an occasional, preliminary to mating, there must be genetically determined variation in the tendency to perform it. Is there any evidence of this? In particular, in view of the theories already discussed, are there heritable differences in the occurrence of conflict behaviour in this, or any other, situation?

Evidence on this point is scanty, mainly because observers have tended to ignore irregular conflict behaviour. The only two significant studies concern differences between species. Hinde in his studies of finches [114], recorded that the canary, greenfinch and goldfinch shared many 'displacement' activities in common and that these appeared regularly also in the hybrids. Displacement breast preening is characteristic of goldfinches alone, and it appears also in goldfinch hybrids but not others; hence this tendency is clearly heritable. Clark, Aronson and Gordon [60] demonstrated heritable differences in pre-courtship behaviour between the platyfish *Xiphophorus maculatus* and the swordtail *Xiphophorus helleri*: the former tends to peck at sand on the bottom of the aquarium, the latter to nibble at the female. F_1 hybrids performed neither of these activities; but in the F_2, thirty-nine per cent pecked and fifteen per cent nibbled, while the backcrosses gave proportions approaching those of the parent types; this strongly suggests a genetical influence.

At present we can say only that, since genetical differences of this type exist between species, they probably also exist within species. But much more work is required.

The mode of gene action

Most of the genetically determined behavioural differences which have been studied seem to be influenced by many genes. A less obvious finding is that the different elements of a complex behaviour pattern may be influenced by quite different genes. This may be true even when they occur in strict sequence and are functionally related. A famous example, although not of courtship behaviour, is worth quoting. It concerns the hygienic activities of honey-bees *Apis mellifera*. Normal bees uncover the cells of diseased larvae and remove the infected insects from the hive. Rothenbuhler, by suitable crosses, obtained colonies which uncovered but did not remove, while others removed only if the cells were first uncovered for them. These two activities seem to be controlled by two independent genes [237, 238]. That courtship sequences are sometimes similarly controlled is suggested by the fact that regular behaviour patterns are sometimes broken up and become unco-ordinated in hybrids.

A third generalization, worthy of a second digression from courtship behaviour, is that certain physiological changes have very specific and unexpected effects upon apparently unrelated behaviour patterns. The larvae of some cultures of meal worm *Ephestia kuhniella* spin a dense mat over the surface of the food before pupating; in other cultures no such behaviour is found. Caspari associated this behaviour with insects of a certain genotype crowded together in food of a certain deficient variety (commercial cornmeal) [59]. It seems that, on this food, insects of the relevant genotype have a retarded development with a long last larval stage. It is during this stage that the larvae come up (in darkness) and crawl around before pupating. They shed silk as they crawl, and the mat is the accidental result of an abnormally long period of crawling. There are several genes involved, all affecting development rate in relation to food: the secondary behavioural effects would scarcely have been suspected without this analysis.

How do genes affect courtship behaviour? A fairly direct way would be to modify sense organs or effectors. There is plenty of evidence that genes can do this in many animals [91] and there is some evidence regarding the repercussions of such changes upon courtship. Changes in sense organs, for instance, can influence a male's ability to keep close to a female. In a series of male fruit-flies, genetically different in eye-pigmentation (and hence visual acuity), mating success varies directly with the degree of pigmentation [92]. Probably flies with inferior sight are unable to follow escaping females, and hence give intermittent displays deficient in the more effective close-range elements. Crossley has evidence that the intermittent displays of ebony flies also result from poor visual reactions [71].

Motor and structural changes may alter the effectiveness of a display movement. For example, small-winged fruit-flies probably fail to fan air effectively towards females [83]. And Pringle found that modifications of effector organs (together with changes in patterns of nervous activity) are associated with the different forms of cicada song [217].

Much more indirect modifications may result from the selection of genes with quite independent effects upon fitness. For example, there may be an increase or decrease of general activity or a modification of habitat preferences, so that the animal has to court in darker or damper or more crowded surroundings. Such changes might immediately modify the display by environmental means. Alternatively they could give advantage to one aspect of a display rather than another, and so promote its exaggeration by further selection. Hence modifications of display will result from the initial change, although they are selected independently. This process is called *secondary selection.*

A probable example of modification by an environmental effect comes from the fruit-fly species *Drosophila melanogaster* and *D. simulans*. It will be remembered that the males of *D. melanogaster* court their very active females mainly on the run and consequently cannot give many of the visual displays typical of *D. simulans* males, for these involve circling and posturing in front of the female. Hence an initial change in the general activity of the two species may have modified male displays, because of their dependence upon behaviour. In fact, Manning has shown that each type of male, given a female with the appropriate behaviour, will court in a manner typical of either species [171] (figure 47).

Another example of female behaviour modifying male display comes from Dilger's parrots [80]. It will be remembered that the switch-sidling display is performed at an intermediate frequency in the hybrids between the two species *Agapornis roseicolis* and *A. fischeri*. But this is only when hybrid males court hybrid females. With females of either parent species, hybrid males display like the males of that species. Whatever the influential factor here (and it has not been analysed), it must be genetically determined, for the hybrid females must possess it to an intermediate degree.

An example of secondary selection (resulting in this case from artificial selection) comes from Ewing's experiments with fruit-flies described on page 133 [83]. His flies, selected for small body size, also had small wings, and he showed that any reduction in wing size (whether by cutting or by genetical means) reduces courtship success (probably because small wings cannot fan air towards the female so effectively). It will be remembered that Ewing found an increased

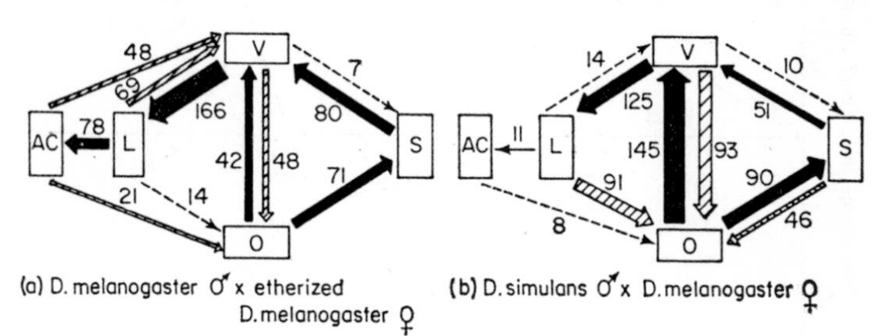

Figure 47. Diagrams showing how *Drosophila melanogaster* and *D. simulans* males may each court in a manner typical of the other, given appropriate females.

(*a*) *D. melanogaster* males courting etherized females (which are inactive like *D. simulans* females) include more scissoring than usual (a typical *D. simulans* feature). (*b*) *D. simulans* males courting *D. melanogaster* females have reduced scissoring (like normal *D. melanogaster* displays). Compare figure 36, page 116. (After Manning [171].)

amount of wing-vibration in the displays of his 'small' strain. But this developed only under certain experimental conditions – when the selected flies were bred together in large numbers. In a parallel experiment in which selected males were paired separately with single selected females, small flies did not show any increase in vibration (figure 48). Evidently the modification was a compensatory one (an increased *amount* of wing fanning compensated for its decreased effectiveness) and it was independently selected in those conditions which made it advantageous when males competed for females. By crossing his two selected lines (large and small), Ewing confirmed that the two characters, body size and vibration frequency, are separately controlled. In the second generation, body size segregated independently of the behavioural change, so that insects appeared with large or small wings together with high or low frequencies of vibration.

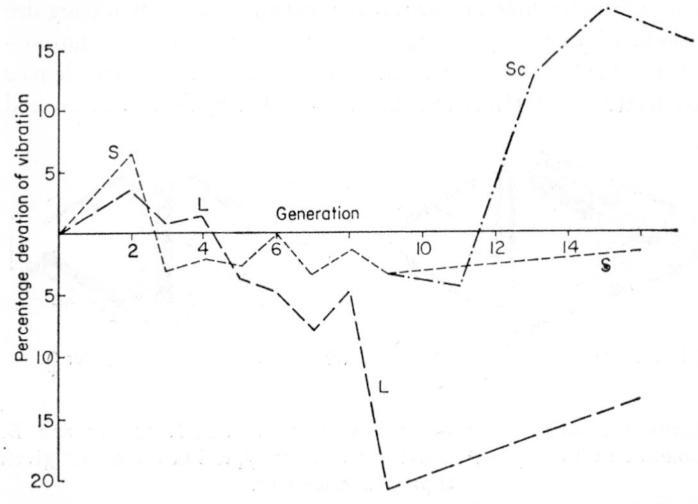

Figure 48. The two populations, L and S, have been selected for large and small body size respectively by a method which eliminates sexual competition between males (single pair matings). The graph plots the deviation of the amount of vibration in their courtship from that shown by control flies. At generation 9 sexual competition is reintroduced (mass matings) for some of the flies of the S population, Sc. The increase in vibration in small flies thus appears to depend on sexual competition.
(After Ewing [83].)

Secondary selection is also suggested by the fact that male displays and female requirements must evolve hand-in-hand. Any change in male display, however determined, is likely to promote a secondary change in female requirements. Alternatively, a change in female perceptiveness could lead to a modification of male display. In the example of *Drosophila melanogaster* and *D. simulans* (if the proposed sequence of events is correct), it might be expected that advantage (in mating more quickly) would go to the *D. melanogaster* females which responded most readily to vibrational stimuli and to the *D. simulans* females which preferred visual ones. In fact the two types of female do seem to differ in this respect. *D. melanogaster* females (unlike *D. simulans*) do not mate readily if their antennae are removed (the antennae perceive vibrational stimuli), while *D. simulans* females (unlike *D. melanogaster*) are not very receptive in the dark [171]. Furthermore, females of inbred populations of 'yellow' fruit-flies are more receptive than are normal flies, as if in compensation for the less effective displays of their males [24]. Male and female behaviour differences are here independent of one another and determined by different genes. And within the large group of fruit-flies many species are sluggish and intermittent in their displays, while others are vigorous and persistent; yet all mate with their own females equally well [253].

Finally there is some evidence that genes might modify courtship by affecting sexual 'motivation'. In fact there is considerable evidence that at least some aspects of sexual responsiveness can be modified very readily by genes, but few investigations indicate how this might alter courtship.

The evidence about genes and 'motivation' comes mostly from selection experiments. Wood-Gush selected domestic cockerels with high and low mating speeds, and obtained a separation of his two strains, with no overlap, as early as the second generation. His 'high speed' cocks were sexually precocious and also adapted more quickly when put with females [293]. Manning similarly selected two strains of *Drosophila melanogaster*. Males of the 'fast' line began to court as soon as they were put into a bottle with females, whereas 'slow' males first did much disturbed running about [172]. In both these experiments, genetical changes seem to have accelerated the initiation of sexual reactions, enabling them to over-ride competing ones more quickly. Rasmussen selected rats *Rattus rattus* for high *versus* low sexual responsiveness: after five generations, rats of the 'high' line would cross an electric grid to reach a female many times more readily than those of the other line [221]. Inbred strains of mice [90, 154] and guinea-pigs [128] also differ consistently in sexual responsiveness; and pairs of monozygous twin cattle *Bos* are very much alike in this respect [15].

These examples probably include more than one type of change, and in no case do we know the physiological basis. One possibility is that genes sometimes affect hormone concentrations, for sexual responsiveness varies with the blood level of sex hormones. This is suggested by the experiments of Wood-Gush and Osborne, who compared pairs of brother fowls [294]. They found differences in the frequencies of certain mating and courtship activities between the pairs, and some of these correlated with comb height (which is often taken as a measure of androgen concentration). Hence differences in comb size and behaviour might both depend upon gene-determined differences in the amount of androgen produced. Sometimes, however, genetically different animals differ in their *responsiveness* to hormones rather than in their hormone concentrations. Goy and Jackaway [96] found that two strains of guinea-pigs *Cavia*, which differed in female receptivity, responded differently to the same massive doses of female sex hormones. Both became more receptive, but the original difference was maintained. Similarly, Bevan *et al.* castrated mice *Mus musculus* of strains which differed in aggressiveness, and then gave them equal doses of male sex hormone. Both strains began to fight again (they stopped when castrated), but the original difference between them could still be seen [33]. Genes, then, can modify sexual responsiveness in at least two ways: by affecting excitatory factors (hormones or sensory inputs) or by modifying the reactivity of the nerve centres to them.

How can changes in sexual responsiveness modify courtship? Few observers have looked for relationships of this sort. However, Manning found that his slow mating lines of fruit-flies had a lower frequency of vibration than his fast lines [172]. Here, then, a frequency change in the patterning of courtship is related to delayed responsiveness to females. In the same species Bastock related the same type of change (a reduction in vibration) to lack of persistence in courtship [24] (figure 49). We do not know whether delayed responsiveness and lack of persistence can be related to the same cause. But one possibility is that they both result from a lowered excitatory state of the neural control system (somehow the 'energizing inputs' have been reduced, or the system as a whole is less responsive). Then a possible explanation of the reduction in fre-

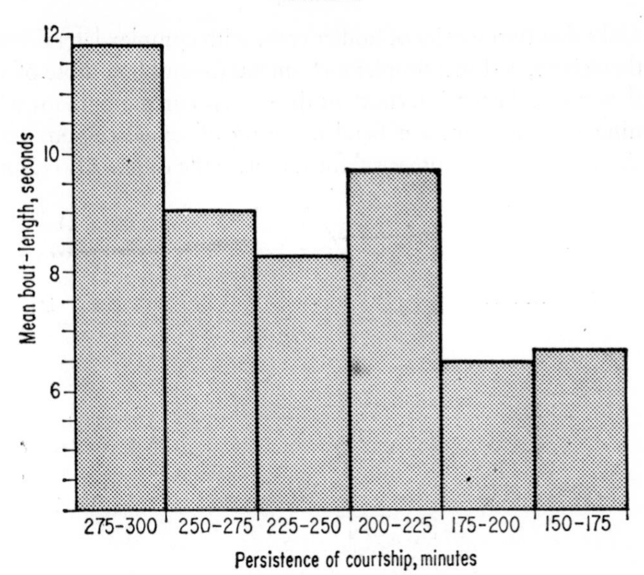

Figure 49. The relationship between the pattern of courtship and its persistence in male fruit-flies *Drosophila melanogaster*, courting females of another species *Drosophila simulans* (which do not accept them). The average length of bouts of vibration and licking in the courtship is plotted against the total time spent courting. There is a significant regression of bout-length upon courtship time.

quency of one element of the courtship pattern but not of another (of vibration, but not of orientation) is that the former requires a higher energy level for its release. Figure 50 illustrates this interpretation for changes found in yellow flies. There are many cases where this type of explanation might be applied. In the guppy or the yellow-hammer, for example, certain activities and displays occur most commonly during early phases of courtship, when the male tends to be less reactive and less persistent in general, while others are confined to later, more intense, stages.

In the same way, changes of input might modify the form of a display. Sometimes there are 'low intensity' and 'high intensity' versions of a display movement which can be correlated with the persistence of sexual reactions as a whole. For example, Crane remarks that two species of fiddler crab, with complex lateral waves of their claws, perform simpler movements (resembling those of vertical wavers) during intermittent displays, occurring perhaps while feeding [68]. Perhaps a general lowering of sexual responsiveness could make the 'low intensity' form replace the other. Also changes

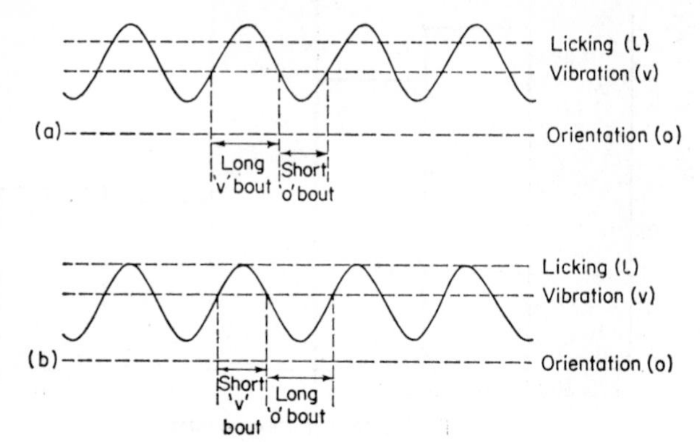

Figure 50. How a change in 'excitatory state' might modify courtship patterns in *Drosophila melanogaster*. The dotted lines represent different energy levels which orientation, vibration and licking might require. Excitation in the control centre is supposed to fluctuate regularly, rising as the male approaches the female, falling when she repels him. High average excitation (*a*) results in a pattern typical of wild type flies (figure 45 A and B); low average excitation (*b*) in a pattern typical of yellow flies (figure 45 C and D).
(After Bastock [24].)

of form like those discussed in gulls and finches concern the relative expression of two reactions in an ambivalent or alternating movement: these could be modified by genetic changes in the responsiveness of either of the control systems. There is evidence that genes can modify responsiveness in attack and escape reactions as well as sexual ones [59]. At present these are no more than theoretical possibilities. Genetic changes in nervous excitability and in thresholds of response may well be responsible for much behavioural variation, but at present evidence is lacking.

To conclude, courtship can be modified by many genes. They may operate by altering sensory or effector organs, by modifying the excitation or the excitability of nervous control systems or by promoting more general changes in form or physiology which have indirect (and often unexpected) repercussions upon behaviour. Many courtship differences between closely related species are genetically determined, and there is some (although scanty) evidence that genetically determined variation of the same nature exists within populations. This variation is mostly in frequency, patterning, timing and occasionally form; hence it could lead to the adaptive modifications discussed in the previous chapter. Often there will have been direct selection for modifications of displays. Any change (environmental or genetical) may open up new possibilities for display. But some changes which increase fitness in quite different ways may modify the effectiveness of a display and hence promote compensatory changes. The study of the genetics of courtship emphasizes once more the interrelationship of all characters.

Part Three: Mechanisms

8: The Sexual Control System

General characteristics

Before a 'mechanism' can be proposed to 'explain' a phenomenon, it is necessary to state clearly the characteristics which it has to account for.

First, all behaviour patterns have a limited duration: they start, proceed for a certain time and then stop. Any control mechanism must therefore incorporate starting and stopping devices (which may be independent of one another). Starting devices are often related to special and highly specific external stimuli sometimes called releasing stimuli; for example, the sound of a bat's high-frequency 'radar' note may cause noctuid moths to take avoiding action [229, 230]. In this case the behaviour appears to be switched on by the stimulus. But some behaviour patterns begin 'spontaneously', with searching or appetitive behaviour whereby the animal appears to seek special stimuli which evoke further reactions: a hungry animal searches for food; a sexually motivated male, for a female. Stopping devices are also sometimes related to external stimuli, but sometimes not. An escaping animal ceases to flee when under 'cover', but it ceases to eat or drink without any new external stimulus to make it do so. Starting and stopping systems must therefore respond to internal states as well as to external stimuli.

Secondly, complex behaviour is usually subject to 'motivational' changes. This term, which was discussed in the introduction, is applied to changes in responsiveness; these may be measured in terms of intensity, duration or elicitability. Many factors may be involved here, operating in many different ways. Most of them appear to be related to changes in the internal state, as when an animal becomes more and more responsive to food as time elapses since the last meal. Sometimes external factors are involved, as in the spring-time increase in sexual responsiveness in seasonal animals. But external factors are rarely immediate in their effect, and often operate by inducing internal chemical changes.

The above generalizations apply to many behaviour patterns. Courtship itself is peculiar in being the preliminary to another activity – mating. Much of the evidence considered in the previous chapters suggests that it occurs when mating behaviour is aroused but opposed. However, its occurrence does, in general, fluctuate together with that of mating itself; it is always difficult to elicit courtship when the mating tendency is low, although mating may sometimes occur without courtship. Few analyses have considered these two activities separately, and it is simplest here to consider the sexual control mechanism as a single entity, bearing in mind that with courtship other (interfering) factors may be involved.

Sexual behaviour usually includes appetitive behaviour: this may be spontaneous or may be induced by specific stimuli such as scents or sounds or flashes of light. The activities of courtship and mating require other stimuli. These are always very specific and usually visual or tactile: they indicate the proximity of the object to be courted. Sexual responsiveness also fluctuates: in particular, responsiveness tends to fall temporarily just after mating. But there is considerable variation here both between species and between the sexes within a species. A male of one species may cease to respond after one mating, another only after several; a male of a given species may recover sexual responsiveness after a few hours, a female of the same species may take many days or weeks. In addition, prolonged courtship (without mating) may result in reduced sexual responsiveness, although it may take a long time to do so.

In most species there are also juvenile and senile periods when sexual responsiveness is low or absent; in many there are seasonal fluctuations and in females, within each breeding season, there may be periodic fluctuations related to readiness to ovulate. Courtship control systems must therefore be expected to differ considerably in detail between species; they may not, however, differ in principle.

Analysis of behaviour

Recently a number of experimental analyses have been made with the aim of obtaining more detailed information about control systems. Usually they consist of experimental interference with factors presumed to be concerned in starting and stopping the activity, and in motivation. Most experiments of this type concern feeding and drinking. This chapter will discuss a few examples relating to courtship and mating.

The first comes from stickleback courtship. This has been investigated by van Iersel [126] and Bol [39]. After nest-building, a male three-spined stickleback gradually increases his sexual responsiveness and maintains it over a fairly long period, usually until he has fertilized the eggs of five successive females.

Sexual responsiveness is measured by the number of zig-zag displays made to a female in a given time, but it is not stable throughout this 'sexual' period. In particular, it falls sharply after each fertilization, recovers quickly on the first four occasions, but remains low on the fifth and finally disappears entirely one or two days later. Bol observed successful courtships and placed a wire loop in the nest entrance immediately after the female had spawned there, thus preventing the male from following her to fertilize the eggs: she found that a fall in sexual responsiveness still occurred. It also occurred when she placed a batch of fresh eggs in the nest of a male which had neither fertilized nor courted. Bol concluded that it is the presence of fresh eggs in the nest (a normal consequence of mating) which reduces responsiveness, not the performance of the sexual activities themselves.

The more complete, more gradual and more persistent decline in responsiveness, normally occurring after the fifth fertilization, seems to depend upon different factors, related to the number of fertilizations. Van Iersel showed that fish which have fertilized fewer than five times also cease to court eventually, if they fail to win further females, but the fewer the fertilizations the longer this takes (figure 51(*a*)). The number of batches of eggs in the nest slightly influences the decay process: more batches tend to accelerate it; but the effect of eggs is negligible compared with that of fertilizations (figure 51(*b*)).

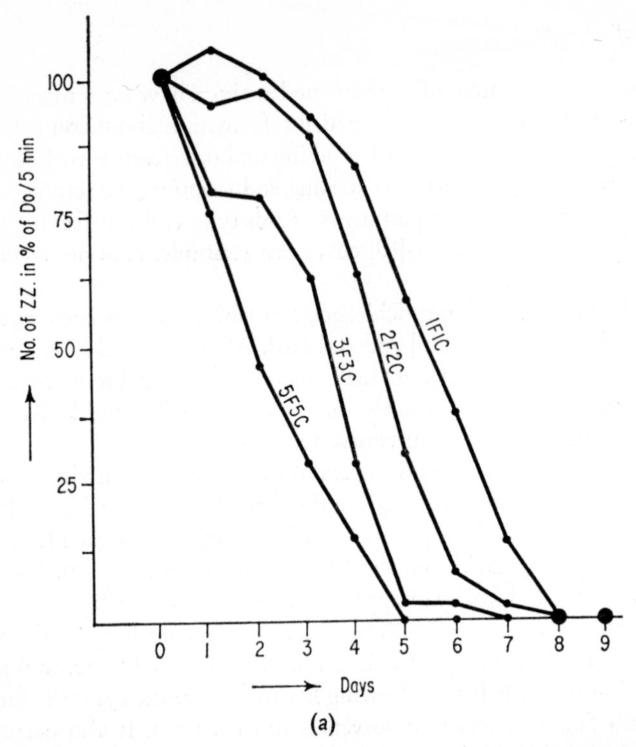

(a)

Figure 51. Graphs illustrating the decline in courtship responses in the male three-spined stickleback after fertilizing different numbers of clutches of eggs. Courting tendency is measured as zig-zags per five-minute test period and expressed as a percentage of the courting tendency on the day of the first fertilization (D_0). F = fertilizations; C = clutches in nest.

Perhaps the factors involved are related to the depletion of sperm; this may be complete after five fertilizations, but sometimes unused sperm may be resorbed slowly. These findings will be discussed after consideration of the next example.

The control of sexual behaviour has been investigated in the females of two species of fruit-fly. Female *Drosophila subobscura* become gradually receptive five or six days after emergence (when their ovaries mature) [223], and remain so until after fertilization:

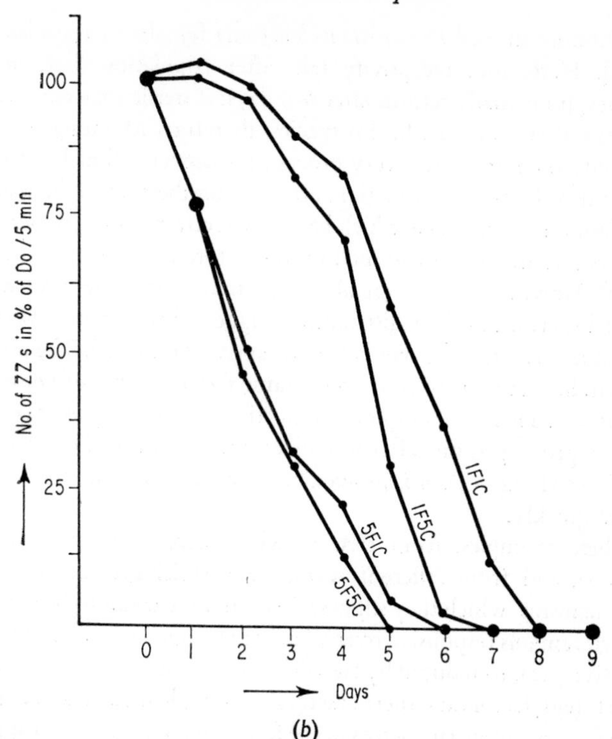

(*b*)

Graph a shows the effect of clutches and fertilizations together (the normal situation). *Graph b* shows the relatively small effect of experimentally adding or removing clutches.
(After van Iersel [126].)

then they become completely unresponsive, usually for the rest of their lives. Maynard Smith [186] experimentally mated females with sterile males (which copulate but pass no sperm): these females recovered their receptivity after 4–24 hours. Hence the immediate fall in responsiveness after every copulation is maintained only if semen is received. Probably, two sets of factors are operating: one is related to copulation, and is rapid and temporary in effect; the other is related to insemination and is long-lasting. The influence of the semen itself is suggested by the fact that it is normally stored in the female throughout her life.

Manning studied *Drosophila melanogaster* females in a similar way [173]. Here, too, receptivity falls after copulation, but, in this species, it normally returns after 6–9 days: if sterile males are used it returns after 12 hours. In this species, therefore, Manning was able to study the normal recovery process. He dissected females at varying intervals after insemination, after noting their sexual behaviour. He found that all insects which had fully recovered their responsiveness before dissection had used up their stored sperm (in fertilizing eggs). However, there were also some unresponsive females which lacked sperm, but their distribution suggested that sperm was used up on the average one day before receptivity returned. The control system here appears to react to the same factors as that of *Drosophila subobscura*. There is a temporary reduction of reactivity after copulation, a prolonged one after insemination. However, the latter does not last so long in *D. melanogaster*, because females use up their sperm more quickly.

These examples, from two widely different groups (fish and insects), and from different sexes, are surprisingly similar in the 'mechanisms' which they suggest. In each there seems to be a system which remains responsive over a fairly prolonged period, but whose reactivity can be reduced by two different sets of factors with distinct effects (egg factors and sperm factors in the male stickleback; copulation factors and sperm factors in the female fruit-fly). One set appears to cause an immediate short-term decline in responsiveness; the other a gradual, prolonged one.

These are the facts. It is probably possible to devise a number of mechanisms and models which behave in this way, but to those who like to suggest 'actual' mechanisms, some of the characteristics are suggestive. In particular, the prolonged periods of unresponsiveness are related in both cases to semen (its absence in the male stickleback and its presence in the female fruit-fly). The factors involved have a slow rate of action and a complete, sometimes final effect. This suggests that their action is chemical, rather than neural; possibly they inactivate the neural mechanisms responsible for the behaviour, perhaps by checking the secretion of hormones. On the other hand the shorter interruptions in responsiveness are in both cases related to specific stimuli (olfactory or visual from fresh eggs; tactile from copulation). These rapid, incomplete, temporary effects may be neural.

A rather different type of analysis is that made by Russell on the mating behaviour of the South African clawed toad *Xenopus laevis* [241]. In captivity, this animal does not mate unless injected with gonadotrophin. When so treated, a male mounts and clasps a female and, if she is responsive, will remain in this position for many hours, until she has spawned and he has fertilized. A female is responsive, if she too has been injected with hormone. Untreated females (and even males) are also clasped, but in this case the male soon dismounts, repeating this behaviour at short intervals. Russell treated males with varying doses of hormone and tested each with the same series of untreated females (in varying order). He found that the total time spent clasping (during a three-hour experiment) varied with the dosage of hormone but not with the females, while the average length of each clasping bout varied with the females and not with the hormone dosage. In other words, factors from the females determined how soon a male would dismount after clasping, but the hormones determined how soon he would remount.

This example agrees with the other two in demonstrating two sets of factors with different effects upon responsiveness. Hormones have a slow-developing but essential effect upon reactiveness (for in their absence the animal is completely unresponsive); they also influence the rate of recovery from the other set of factors. These, coming from unresponsive animals, are in the form of sensory stimuli (Russell believes them to be sounds uttered by unwilling but not by willing animals). Again they are rapid and temporary in their effect.

From these three examples (although they are lamentably few) it is possible to outline some of the characteristics of a sexual control mechanism. We may assume that it is a neural mechanism since all behaviour in higher animals is controlled by the nervous system. This mechanism is capable of reacting to specific stimuli (from the female) and of organizing courtship behaviour in response. Its responsiveness to these stimuli can be modified by chemical factors (by concentrations of hormones or possibly by agents from semen). If these factors remain stable, the system may remain in a continuous state of responsiveness or non-responsiveness over a long period.

Sensory stimuli may also modify responsiveness more immediately and more temporarily. This type of control system, incorporating both short- and long-term switch-offs, may be a very common one: it has been demonstrated, for example, for feeding in both insects and mammals. The operating factors will vary in every case. The sexual system is likely to be elucidated further by studies of hormonal mechanisms on the one hand, and neural mechanisms on the other. The next two chapters will deal with each in turn.

9: Hormonal Mechanisms

This chapter investigates the role of hormones as energizers of sexual behaviour. It discusses the effects and the interactions of gonad and pituitary hormones in vertebrates. Hormones certainly exist in invertebrates too, and may play a similar role; but the evidence here is incomplete: vertebrate investigations provide a clearer picture of their relationship to the sexual control system.

Gonad hormones and courtship

There have been many demonstrations that sexual behaviour in vertebrates is influenced by hormones secreted by the gonads [53, 54]. When there is a breeding season or a limited period of maturity, breeding behaviour coincides with the presence of large, functional gonads actively secreting hormones (androgens from the testis and oestrogens from the ovary). When these hormones disappear for any reason, because of castration or disease or senility, or because of seasonal regression of the gonads, the behaviour usually disappears too. The gonads of seasonal animals are always large during the breeding season and small at other times. Males usually have a correspondingly prolonged and continuous period of sexual activity followed by a complete absence of response. Bullough [52] has even demonstrated in British starlings *Sturnus vulgaris britannicus* a detailed correlation between testis size and the intensity of sexual behaviour: here testis enlargement (and breeding behaviour) begins in September and October and continues at a low level until spring, when there is a sudden acceleration. This is in marked contrast to the migratory continental starling *Sturnus vulgaris vulgaris*, in which Kluijver has shown that gonad growth is delayed until February and breeding behaviour does not occur until March [134]. Occasionally sporadic out-of-season breeding behaviour does occur, particularly in birds during a mild autumn when spring-like renewals of song and courting may be observed. It is then usually found that the gonads have temporarily enlarged after their normal regression.

In contrast to males, females are not continuously ready to behave sexually during the breeding season. Their hormone production and behaviour is closely linked to the presence of ripe follicles in the ovary and this occurs often only at intervals. In mammals in parti-

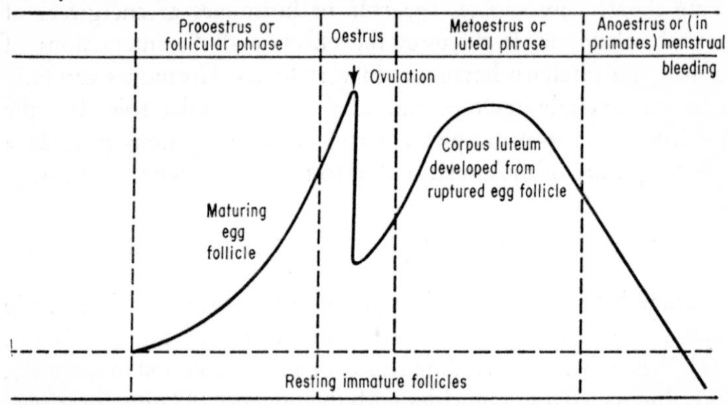

Figure 52. **Diagram illustrating the events in the ovary during a single mammalian oestrus cycle. After ovulation the ruptured egg follicle becomes a secretory body, the corpus luteum: this is maintained if pregnancy occurs and does not regress as shown in the diagram. In the absence of pregnancy, a new cycle may start after a short anoestrous (or bleeding period in primates) or there may be a prolonged pause. (Modified from Bullough [54].)**

cular, there is usually a recurrent cycle of activity in the ovary, the oestrous cycle (see figure 52). Some mammals have only one cycle in each breeding season, but most are polyoestrous. In the absence of fertilization, cycles succeed one another at regular intervals; after four days in the laboratory rat, four weeks in women, and four months in the domestic cat (these three species have no breeding season). There are two variants upon this cycle. In the rat, no corpus luteum is formed unless copulation occurs. In its absence oestrus is quickly succeeded by the next proestrus. More commonly (in the cat *Felis*, rabbit *Oryctolagus*, ferret *Mustela furo*, ground squirrel *Citellus*, shrew *Sorex* and mink *Mustela lutreola*), ovulation itself occurs only after copulation. Failing this, oestrus slowly changes to anoestrus and a new cycle follows in due course.

The correlation of female sexual behaviour with the oestrous phase of the cycle is easy in many mammals, because all the phases can be judged by an examination of cells scraped from the wall of the vagina. The peak of sexual receptivity of almost all mammals occurs at oestrus (during the peak of oestrogen secretion), and usually females permit copulation only at this time. They are then said to be 'on heat'. Oestrous females become restless and 'nervous'; they seek out males and crouch or otherwise invite coitus. Oestrous female lions *Panthera leo* head-rub, crawl under other lions and make characteristic sexual calls [61]; mares neigh, urinate frequently and approach stallions; heifers are restless, bellow and ride other cattle; sows *Sus* eat sparingly [225]; golden hamster females *Cricetus* lose the aggression they normally show to males [132]. Oestrous rats are more willing to cross an electric grid to reach a male than at other times [258], and Richter showed that they also have a peak of running activity at this time [227].

Experiments involving castration and the injection of hormones are also impressive [31, 94]. Removal of the gonads nearly always results in the loss of sexual behaviour, although these effects are more immediate and complete in females than in males. Nearly all female mammals lose their receptivity at once, whether the ovaries are removed before or after maturity (except occasionally in adult chimpanzees and women). As for males, the same result usually follows if the testes are removed before maturity, although here the rat *Rattus*, hamster *Cricetus* and guinea-pig *Cavia* (all rodents) are exceptional. However, the castration of adult males has very varied consequences. There is often only partial and gradual loss of sexual activity: examples include some fish (the jewel fish and the fighting fish), some birds (the cock and the pigeon *Columba*), and many mammals (especially bulls, horses, cats and primates).

The restoration of sexual activity by the injection or implantation of hormones or gonadial tissue is almost always effective: it has been demonstrated in castrates, young or out-of-season animals, even senile or sick animals, and those genetically impotent. Beach even restored sexual behaviour to a rat which had lost it after damage to the cerebral cortex [28, 31]. The only exceptions are among human beings, where results vary.

In all these experiments the amount of hormone injected is important. Sometimes certain components of a complex pattern of sexual behaviour require more hormone than others. After castration rats lose the ejaculatory pattern first, and regain it last after hormone treatment [257]. Perhaps the most precise experiment demonstrating the relationship between hormone dosage and sexual behaviour was performed by Russell on the clawed toad *Xenopus laevis* in the experiments described earlier: the total time which a male spent clasping an unreceptive female varied directly with hormone dosage [241].

In general then, sexual behaviour is very considerably affected by gonad hormones, often to the extent that it cannot occur in their absence. However, to understand the characteristics of hormone action, it is important to consider the exceptional cases too. The most notable exceptions occur amongst the primates and man. Man in particular varies greatly in his dependence upon hormones. The menstrual cycle of women may be regarded as equivalent to the mammalian oestrous cycle in its dependence upon hormones. But there is little correlation of behaviour. Women remain receptive throughout the menstrual cycle and although they may have peaks of receptivity at certain phases, these do not necessarily coincide with peaks of oestrogen secretion. People of both sexes may retain sexual activity after their gonads have been removed, or have ceased to function for any reason, and both vary in their responsiveness to hormones. The same is true of many other primates. It may be that the old control centres of sexual behaviour have become emancipated from hormone influence in these animals; but it is more likely that new factors (perhaps new nervous influences from the brain) have come to have increasing effect.

It was originally thought that emancipation from hormone influence was an evolutionary trend reaching its culmination in animals with the greatest development of control centres in the cerebral cortex. However, many of the other exceptional cases, particularly those, already quoted, among lower male mammals, remain to be explained. The anomalies may sometimes be due to the secretion of androgenic hormones, similar to the testis hormones from the adrenal glands. However, there is increasing evidence, particularly among mammals, that experience plays an important role both in organizing the development of sexual behaviour patterns and in determining their later dependence upon hormones. Two examples are worth quoting.

The male domestic cat *Felis catus* has no sexual behaviour until puberty: evidently it depends completely upon sex hormones. However, experience after puberty may modify this dependence. Castration after such experience results in a very gradual decline in sexual responsiveness, some activity being retained over a very long period. By contrast, castration of cats of similar age, but lacking sexual experience, results in an immediate loss of all sexual behaviour [233].

The second example comes from the male guinea-pig *Cavia* [97, 276, 277, 278, 296]. Here sexual play occurs in infancy, apparently uninfluenced by sexual hormones (it can occur in animals castrated at birth). This experience may influence later responsiveness to hormones. Guinea-pigs deprived of such experience, by being isolated immediately after weaning, are always less sexually active when they come to maturity. This cannot be explained by hormone deficiency, for the same differences are seen if the animals are castrated after maturity and given identical doses of hormone. All lose their sexual behaviour about 10 weeks after castration, but the isolated animals return to a lower level of responsiveness after treatment. Here there is no change in dependence upon hormones, but only on responsiveness to them. Perhaps the early experience facilitated the development of function in the control centres of the brain. However, this experiment does not rule out the possibility that effects of isolation other than lack of sexual play are responsible.

Genetical differences, too, can influence responsiveness. Grunt and Young [100] distinguished three groups of guinea-pigs, with high, medium and low sexual activity, respectively. After castration and hormone treatment the differences remained. Work on these anomalous cases is continuing and is still incomplete, but it is obvious that hormones are not the only factors concerned in the activation of sexual control mechanisms.

A final point should be made regarding sexual specificity. It might be supposed that mechanisms for male behaviour exist only in males and those for female patterns in females. Similarly, it might be supposed that testosterones induce only male behaviour, and oestrogens, only female. But neither statement is true. Members of each sex can perform the behaviour appropriate to the other, if treated when young or castrated with large doses of the 'opposite' hormones. Spayed females treated with testosterone behave like males, and castrated males, dosed with oestrogen, behave like females. Usually the dosage required is very large, and the behaviour is often incomplete and less vigorous than normal; nevertheless it is clearly recognizable.

Sometimes, however, oestrogen injected in large quantities into castrated male animals will induce, not female, but male behaviour. Here it is the action of the hormone which is unspecific. It is also possible, although more difficult, to induce female behaviour with testosterone in spayed females. It has even been stated that androgen is more effective than oestrogen in restoring sexual behaviour in women after menopause. Occasionally, both male and female behaviour are induced together when either hormone is used, and there are reports of treated animals behaving like males when with females and like females when with males. Conclusions about the non-specificity of hormones are, however, open to some question, because the doses involved are so large that there may be secondary effects.

In the normal animal, then, a nervous organization develops, capable of promoting the sexual behaviour of both sexes. Usually the appropriate hormones act as selectors, activating only one of these alternative control systems. However, there is a bias in favour of the 'correct' control system in each sex; female systems are more readily energized by oestrogen and male systems by testosterone. And each hormone, besides strongly activating its 'own' type of behaviour may, in addition, weakly activate the opposite one [30].

The pituitary: the link with the environment

The account of the link between gonad hormones and sexual behaviour in vertebrates can be carried one stage further. It is one thing to demonstrate that sexual behaviour and gonad activity fluctuate together, but another to understand how these fluctuations are synchronized with annual seasonal changes. It is unlikely that the gonads could be directly responsive to seasonal changes: there must be another means of control; and this has now been traced to the pituitary [54]. This gland lies beneath the brain and is connected to the hypothalamus in the floor of the fore-brain by both nerves and blood vessels (figure 59, page 178). It is thus in a good position to be influenced by the brain and hence by the sense organs. The anterior lobe, like the gonads, changes in size and activity in step with breeding behaviour [190, 220, 284, 290]. Moreover, gonadotrophic hormones, secreted by it, can induce sexual behaviour in immature or out-of-season animals. However, they are ineffective in castrated animals. Evidently their action is indirect: they stimulate the gonads and the gonad hormones activate the behavioural control systems in the brain.

In mammals the pituitary influences the gonads by means of two hormones, follicle-stimulating hormone (FSH) and luteinizing hormone (LH). The latter is sometimes known as interstitial cell-stimulating hormone (ICSH) because it promotes the growth of the structural or interstitial cells of the gonads. FSH, by contrast, promotes the development of the germ cells, stimulating spermatogenesis in the testes and the growth of ova and ovarian follicles in the ovaries. These two hormones increase in quantity at the beginning of each breeding season and are withdrawn at the end of it. In addition they control the oestrous cycle in female mammals. FSH is dominant during the first phase of this cycle, promoting follicle development and oestrogen secretion (from the follicle walls). As oestrogen reaches a peak at oestrus it inhibits FSH, so that LH replaces it as the dominant pituitary hormone. This first induces ovulation and then stimulates the formation of the corpus luteum which secretes progesterone. The latter co-operates with oestrogen in continuing to suppress FSH, so that it is only when the corpus luteum degenerates, and both progesterone and oestrogen are at a low level, that a new cycle can begin (figure 53). During pregnancy, hormones derived from the placenta (chorionic gonadotrophins) maintain the corpus luteum and so prevent further cycles. It is these placental hormones, which are also found in urine during pregnancy, which are used as a diagnostic test for pregnancy in women. When

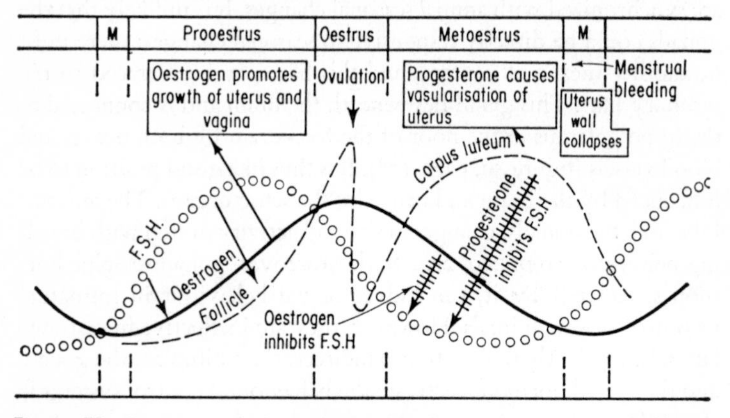

Figure 53. Diagram of the relationship between pituitary FSH and oestrogen during the menstrual cycle of women. The plain arrows indicate stimulating influences of the hormones concerned: the hatched arrows, inhibiting influences. LH (not shown) alternates with FSH and maintains the corpus luteum which secretes progesterone (also not shown). All hormones are at a low concentration when menstrual bleeding occurs.
(Modified from Corner [62].)

present, they induce ovulation in the female clawed toad *Xenopus laevis*, which does not normally ovulate in captivity [117].

Males have no cycle of activity, although testosterone, thought to be secreted by the interstitial cells of the testis, does, like oestrogen, inhibit FSH production. Instead, the three hormones, testosterone, FSH and LH, tend to strike a balance during the breeding season instead of replacing one another cyclically. The reason for this difference is unknown, but male and female pituitaries appear to differ for, after the gonads have been removed, FSH production can be suppressed in female pituitaries only by ovarian implants, and in male pituitaries only by testis implants. Implants (and hence hormones) of the opposite sex are not effective [211, 212]. However, these differences may themselves be due to differential influence from the hypothalamus, for there is evidence, summarized by Harris, that this region of the brain develops different types of activity with respect to the pituitary in the two sexes very early in life [105].

Much is now known of the way in which the pituitary can be influenced by the brain [99, 105]. Communication is achieved, not through neural connections as might be expected, but through the large blood vessels which run through the hypothalamus stalk (figure 54). These vessels arise from a capillary network, on the ventral surface of the hypothalamus, which is in close contact with a

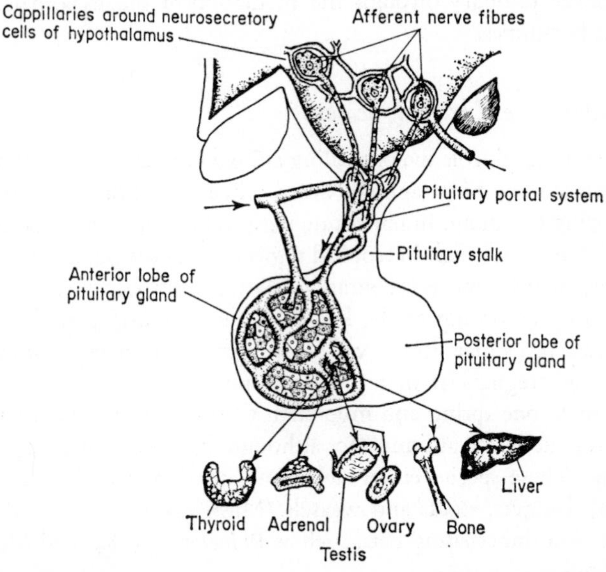

Figure 54. Neurosecretory control of the anterior lobe of the pituitary. Three neurosecretory cells are shown in the floor of the hypothalamus above. Their axons run down the pituitary stalk and pass their secretion into capillaries of the pituitary portal system. These carry the secretion to secretory cells of the anterior lobe of the pituitary and stimulate them to pass their special hormones into the blood stream. Some of these influence the gonads. (After Netter, from Etkin [82].)

plexus of nerve fibres looping continuously into the deeper layers of the brain. It is believed that the nerve plexus is neurosecretory, and that the blood vessels carry hormones which it produces on stimulation. These influence the pituitary. Harris has shown that electrical stimulation of the hypothalamus of mammals can result in the secretion of pituitary hormones and that interruption of the blood vessels interrupts the oestrous cycle [105]. Further, stimuli which can induce testis growth in male birds (for instance increased illumination) are ineffective when these blood vessels are cut. No such effects result from the cutting of the neural connections which also link the pituitary and the hypothalamus. Evidently, then, specific external changes (presumably acting on the sense organs) can influence the pituitary through the mediation of the hypothalamus and its hormones.

The pituitary and breeding seasons

The pituitary's role in controlling animal breeding seasons is now well established. Most animals breed when conditions are best for the rearing of young. In temperate regions the young must be born in the spring, although in tropical regions a wet season, at any time of year, may be more favourable. Mating must occur some time before this, according to the length of the gestation period. Most temperate birds mate in the very early spring, but many mammals with long pregnancies mate in the autumn or winter [5]. Horses *Equus* mate one spring and give birth the next. Many hibernating animals mate in the autumn too, although their pregnancies are not so long. These species either delay implantation (roe deer *Capreolus* [5], badgers, stoats and weasels (Mustelidae) [104]), or delay fertilization (hibernating bats, such as *Rhinolophus* [184] and *Myotis* [5]).

If the pituitary is to determine the season of mating by responding to environmental changes, it must react to very different stimuli in different species, or else react to the same stimuli in different ways [54, 56, 178, 180, 181, 182, 183, 239]. In fact, most spring-breeding vertebrates react positively to light and warmth, and negatively to cold and reduced light. The light response is to gradual changes occurring over a long period; sudden prolonged increases or decreases provided experimentally are effective, but less so than the natural condition. Autumn breeders, by contrast, respond to these same factors, but in the opposite sense. Of these two factors, light is usually more important than temperature: this is not surprising, for it is the only seasonal environmental change which occurs entirely regularly. Many animals, particularly birds, need to regulate their mating period very precisely, for they may rear two or three broods before the autumn, and to do this they must mate by a certain date even though the weather remains cold and wintry. Other animals respond to such features as rainfall, increased food supply or increased nesting material. The North American toad *Scaphiopus bombifrons* mates only after the first heavy rainfall of mid-spring [12], and the African weaver bird *Quelea quelea* responds to the increased nesting material which appears suddenly after heavy rains[180]. In this species, experienced (but not young) birds respond to the rainfall itself. This must be a learned response. The regulating factors for yet other species are unknown. Many of the animals of the New Hebrides, for example, maintain very precise breeding seasons, although there is little variation of temperature, day-length or rainfall throughout the year [12, 13].

The pituitary also possesses autonomous rhythms, although these are normally obscured by the influence of external factors. Marshall finds that the testes of many birds (probably under the influence of the pituitary) become refractory immediately after the breeding season, failing to respond to any of the stimuli which normally induce their growth [178, 180]. Responsiveness is usually restored by the autumn, but is soon inhibited by cold and decreasing light. This explains the temporary outbursts of sexual behaviour common among birds in a mild autumn (which may even proceed as far as egg-laying). Some of the birds of Ascension Island are peculiar in breeding every nine months under almost uniform conditions. Marshall suggests that this too represents an autonomous pituitary cycle, albeit an unusually long one.

The pituitary: the gonads and courtship

More recently, other external influences, operating mainly upon the pituitaries of females, have been revealed. These are of considerable importance in synchronizing the sexual responsiveness of male and female; often quite elaborate mechanisms ensure that each is ready to react in the right way at the right time. This has been particularly worked out for birds. In many species seasonal environmental factors are, by themselves, insufficient to bring about the full spring-time maturation of the female gonads [23]. Hence, at the beginning of each breeding season, the males become sexually active before the females. They leave the winter flocks first to set up territories and, in migratory species, they arrive first at the breeding areas. And even when they do arrive, the females are not yet ready to mate. It seems that the final spurt of ovarian growth occurs only in the presence of the male. Craig demonstrated this in pigeons *Columba* over fifty years ago, when he showed that females will not ovulate when solitary but do so if they can see males and spend some time with them daily [63]. Later, Matthews found that the latter requirement was unnecessary, and the sight and sound of the males alone had the same effect [185]. And recently Erickson and Lehrman showed that it was the courtship of the males which was significant; castrated males (which do not court) do not induce ovulation when shown to females [149]. Polikarpova subjected out-of-season female house-sparrows *Passer domesticus* to increased illumination in two groups, one with males and one without; only the first group developed mature ovaries [213]. Burger, in a similar experiment, tested three groups of starlings *Sturnus vulgaris*: (i) isolated or grouped females, (ii) grouped males and females, (iii) males and females in single pairs. In the first group the oocytes grew to 3 mm, in the second to 5 mm, and in the third to 10 mm [55]. This experiment indicates that some degree of isolation for pairs of birds may be significant, and suggests that courtship plays a part in the maturation process.

Besides courtship, nest-building appears to be important in the pattern of female development. Usually it follows an initial period of courtship, coincides with the first signs of female receptivity, and

Figure 55. The sequence of events when a pair of ring doves *Streptopelia risoria* are put together in a place provided with nesting facilities. (a) Introduction; (*b*) courtship; (*c*) selection of nest site; (*d*) building; (*e*) egg-laying and brooding. Courtship overlaps with (*d*) and (*c*); mating occurs mainly during (*d*).
(After Lehrman [149].)

stimulates ovulation a few days later. The female snow-bunting *Plectrophenax nivalis subnivalis*, according to Tinbergen, becomes receptive after about two or three weeks' courtship and begins to nest-build at the same time [267]. Copulation, too, is often limited to the nest-building period. Mating and nest-building are usually followed closely by egg-laying; and most female birds (with the notable exception of the gallinaceous ducks (Anatidae), grouse (Tetraonidae) and poultry *Gallus*) will not lay eggs in the absence of a mate. Hence a closely knit sequence – courting, mating and nest-building, followed by egg-laying and brooding – occurs in most birds, and this is largely determined by pituitary responses to stimuli received during courtship and nest-building.

Lehrman, working with ring doves *Streptopelia risoria*, is the first to suggest the hormonal mechanisms underlying this process [149, 150]. Solitary females fail to respond either to nest material or to eggs (by brooding), although they very occasionally lay eggs. A pair, however, first courts for one to three days, builds a nest and mates during the following week, then lays eggs and broods one to three days later (figure 55). Apparently courtship is a necessary preliminary to nest-building, and nest-building to brooding. This is demonstrated by the experiments illustrated in figure 56. However, each of these stages can be dispensed with by appropriate treatment with hormones. A bird which is first treated with oestrogen and then given mate, nest and eggs, behaves as if it has courted; it builds first and then sits one to three days later. If it is treated with progesterone,

---

Figure 56. Experiments showing that the sequence courtship, nest-building, brooding, in the ring dove cannot normally be short circuited. Pairs were given a complete nest and eggs on day 0 after various previous treatment. Histograms indicate the proportion of pairs which brooded on given days afterwards. (After Lehrman [149].)

(1) No previous association: courting and building before incubating.
(2) As above, but birds kept separately in cage first (to test for habituation): courting and building before incubation.
(3) Courting and building allowed previously: brooding immediate.
(4) Courting only, allowed previously (nest material absent): building first, then incubating.

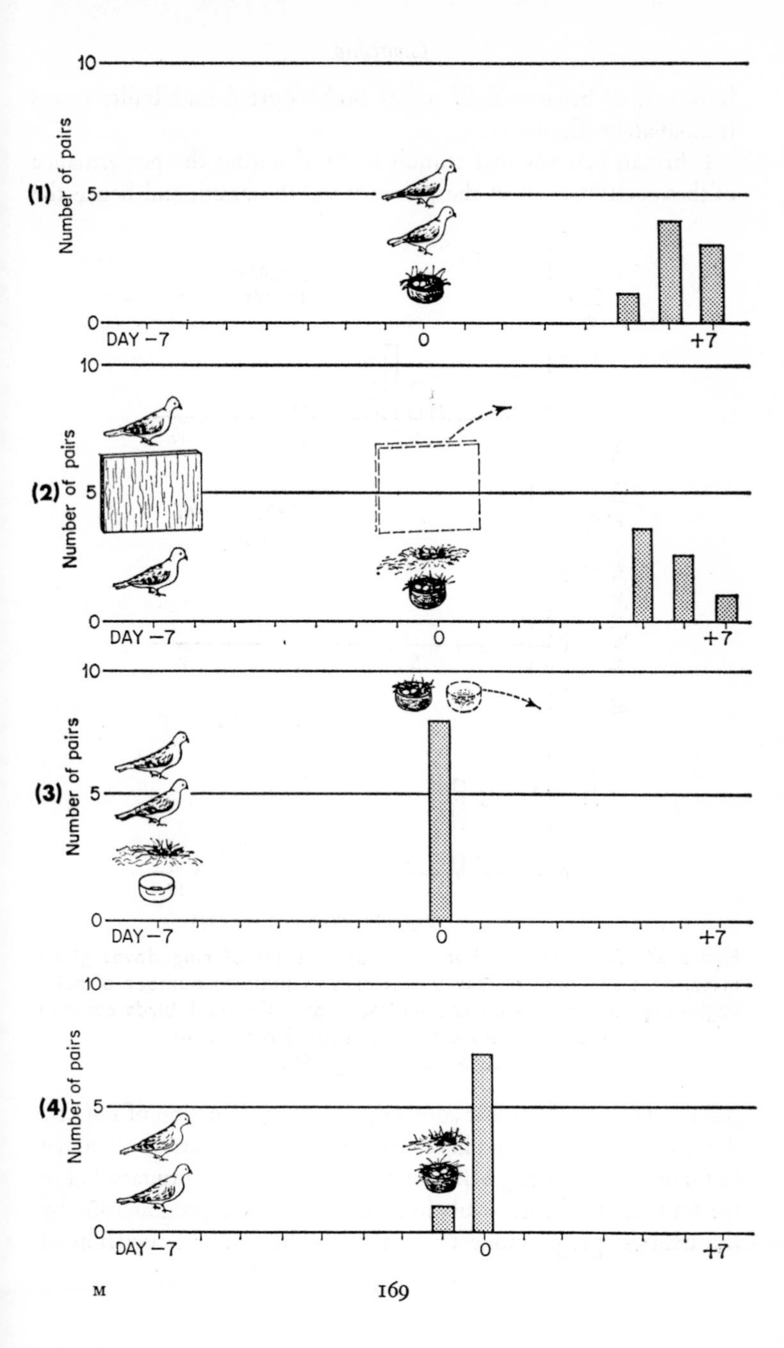

10
Number of pairs
(1)
5
0
DAY −7
0
+7
10
Number of pairs
(2)
5
0
DAY −7
0
+7
10
Number of pairs
(3)
5
0
DAY −7
0
+7
10
Number of pairs
(4)
5
0
DAY −7
0
+7

however, it behaves as if it has both courted and built; it sits immediately (figure 57).

Lehrman believes that stimuli received during the performance of these activities affect the pituitary via the brain, and hence the

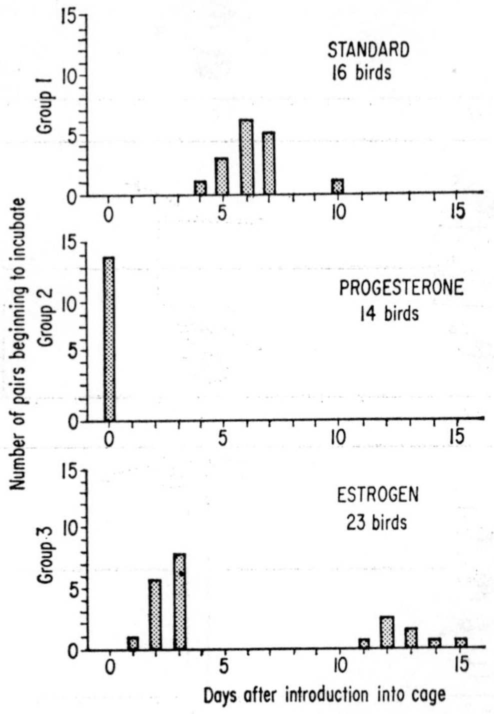

Figure 57. The effect of hormones upon pairs of ring doves given mate, nest and eggs on day 0. Histograms indicate number of pairs beginning to incubate on the indicated day. Group 1 birds courted and built before sitting; group 3 birds built.
(After Lehrman [145].)

gonads. He postulates that courtship induces production of FSH by the female's pituitary and hence oestrogen by the ovaries; oestrogen facilitates nest-building behaviour, which promotes pituitary LH in the female, and hence ovulation and secretion of progesterone by the ovaries; progesterone facilitates brooding. The association of ovulation with courtship and nest-building is shown in figure 58. To complete the story, brooding in its turn promotes prolactin production from the pituitary, which facilitates the feeding of the young. This sequence has to be initiated by male courtship, and the male's pituitary and gonads must therefore be actively secreting beforehand, so that, primed by testosterone, he is immediately

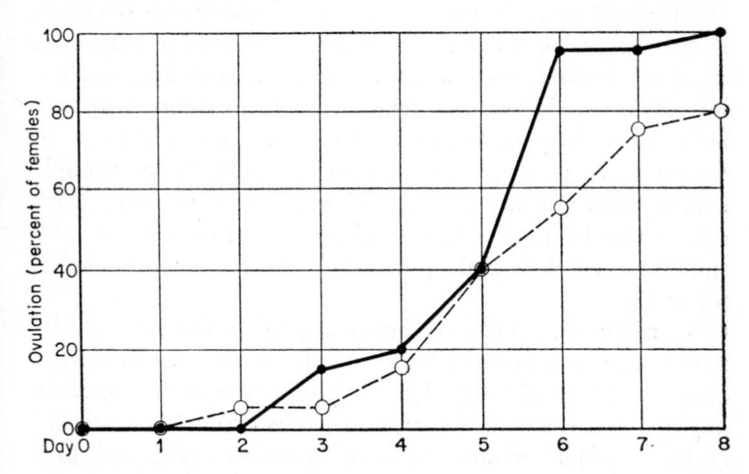

Figure 58. Ovulation in ring doves as a function of times previously spent with mate (open circles) or with mate and nesting material (solid dots). The abscissa gives length of association for different groups of birds. The plotted points show what percentage of each group ovulated on given days since association. (After Lehrman [149].)

ready to court. He, too, has a cycle leading up to parental behaviour. Apparently nest-building behaviour brings him into a physiological condition in which brooding (or the sight of the female brooding) will induce prolactin secretion and hence parental feeding.

Synchronizing processes like these probably occur in many species of birds. They require an association of the pair, before mating, of at least some days, and this occurs in all but a few species. In addition, communal displays, chases and flights in the early part of the breeding season may also stimulate female pituitaries and help promote synchrony of breeding in communal birds. Few other vertebrates (apart from a few fish and reptiles) require this strict sequence of mating, nest-building and brooding; nevertheless there is some support for the idea that male courtship does often stimulate female ovulation. Most fish will spawn in the absence of males, but male courtship (and sometimes even the sight of males) may increase its frequency and the number of eggs laid [4]. Clasping by male frogs and toads (Anura) changes the pattern of spontaneous ovulation, from a gradual sporadic extrusion to a precise co-ordinated one [201]. It is not known whether any of these effects are brought about by pituitary influence, but females are often exposed to male influences for long enough for this to be so. Clasping by frogs and toads may be prolonged; and, among sticklebacks, for example, females often visit many male territories before spawning. These devices may be important in ensuring that ovulation occurs at the right time and place, in the presence of a male and in a good place for laying.

For mammals, which have internal gestation, the needs are different. Ovulation needs to be regulated, not so much to external events, as to internal ones, like the preparation of the uterus for implantation. The ovulation of most species occurs 'automatically' in the rhythmic pituitary-regulated oestrous cycle, and male behaviour must be synchronized with it rather than *vice versa*. Normally this is achieved by the male's sexual responses to the scent and behaviour of the female when on heat. However, there is some evidence of male influence upon female cycles, especially at the start of the breeding season. Often male courtship accelerates the first oestrus; mink *Mustela lutreola* breeders deliberately put females with males early in the season for this reason [30]. In other species, the presence of males appears to accelerate the early stages of any oestrous cycle. This has been demonstrated in cattle [219] and mice (in the latter, even in the absence of copulation). House mice *Mus musculus* may react to the smell of strangers of their own species with a failure of the oestrous cycle: this can be overcome by the presence of a male which courts and mates with them [285, 286]. A large group of females, strangers to each other, remain in anoestrus unless a male is introduced; then they all come into oestrus simultaneously three days later. (Females reared together from weaning do not interfere with each other's cycles in this way [286].) The smell of a strange male can, under certain conditions, have another effect. If a recently inseminated female house mouse or deer mouse (*Peromyscus*) has her mate removed and replaced by a strange male, implantation of the ova may be prevented: the female returns to oestrus as if no mating had occurred [49, 50, 206]. Evidently the smell of strangers can interfere with various stages of the female cycle, but male smells always induce oestrus.

Then there are the mammals (mentioned on page 157) which ovulate only after copulation. These animals have cycles in other respects just like those of the automatic ovulators. The influence of the pituitary has been demonstrated in the rabbit *Oryctolagus* [106, 175]. Ovulation normally occurs about ten hours after copulation, and removal of the pituitary within one hour of this event (but not later) can prevent it. Presumably by this time the gland has already secreted the hormone which induces ovulation. A particular type of cell in the pituitary, thought to be concerned with LH secretion, increases rapidly after coitus. Ovulation can also be induced by electrical stimulation of the hypothalamus, and severing of the stalk connecting this region with the pituitary prevents this. This suggests that in these species the pituitary induces ovulation (by secreting LH) only when it is itself stimulated by the hypothalamus. Presumably sensory stimuli received during copulation are in some way fed to the hypothalamus.

All these hitherto unsuspected adjustments to mammalian cycles suggest that these are far less automatic than had been assumed. Usually the adjustments are such that conditions suitable for mating bring a female rapidly into a state in which fertilization can occur.

Hormones and behavioural mechanisms

There emerges a picture of interacting systems. The gonad hormones evoke sexual behaviour, presumably by acting on nervous control centres. But hormone secretion can be modified in response to external stimuli through the pituitary and the hypothalamus. Sexual responsiveness can thus be synchronized with changes in the environment. And it can be synchronized with physiological changes in the organs of reproduction too, because hormones can influence both simultaneously.

But how do the hormones act on the systems which control the behaviour? There is more than one possibility [297]. They could act directly on the nervous system, or they could have an indirect effect by an action elsewhere in the body.

One example of a possible indirect effect comes from the regurgitation behaviour of ring doves *Streptopelia risoria* [143, 147, 150]. The hormone prolactin causes glands in the crop wall to secrete 'milk' which gradually distends that organ. Often the squabs in the nest induce regurgitation by mechanically disturbing the distended crop while the parent is sitting, thus producing food for themselves. However, experienced birds tend to approach squabs once their crops are distended, and regurgitate on sight or sound of them. But they do not do so if the skin in the crop region has been anaesthetized. It seems that stimuli from the distended crop are responsible for the appearance of this appetitive behaviour (which is in this case influenced by experience). Here the brain is acted on by sensory inputs, resulting from hormone action. The hormones of brooding birds, too, increase the blood supply to brood patches; hence temperature receptors are stimulated in the skin, and their input to the brain may set going the neural processes concerned in sitting on the eggs.

Hormones may also bring about chemical or physical changes which could influence brain centres. For instance, pituitary hormones are said to cause osmoregulatory changes in some marine sticklebacks which coincide with their migration into fresh water [34, 135]. It is possible that some chemical outcome of these changes could be involved.

Courtship and mating, too, may be influenced by indirect effects of hormones. Gonadial hormones undoubtedly have peripheral effects, particularly in mammals [297]. They increase the sensitivity of the skin in the genital region of both males and females; they also increase sensitivity to specific scents. Women can smell musk and urinary steroids after puberty (but not if their ovaries are removed) [151, 153], and male white rats can distinguish a specific female odour (but not after castration) [152]. Such effects could provide sensory inputs to the central nervous system which might lower the threshold for mating. However, their influence is not essential, for oestrous behaviour in female mammals is not suppressed by the removal or denervation of the genital tract [106, 14, 17, 45, 232]. It is very likely that sensory stimuli from these regions are fed into the central nervous system, but they are not major factors in sexual behaviour. Perhaps they are more important in initiating appetitive behaviour than in evoking the behaviour pattern itself. This may also be true of the ring dove *Streptopelia risoria* for in a related species *Streptopelia roseogrisea* regurgitation itself (as distinct from appetitive searching) can be induced by young squabs presented only in the incubation period, before the parent crop has engorged [133].

What is the evidence for a direct effect of hormones upon neural centres? For a long time it was questioned whether hormones could act in this way, because of the specificity of their effects. It was believed that all nerve cells were basically similar in chemistry and physiology, and it was difficult to imagine how hormones could affect some and not others. However, great physiological differences are now known between different types of neuron. More important still, special regions of the hypothalamus are found to respond to changes in the concentrations in the blood of carbon dioxide, urea, and food products, and to dehydration and alterations of temperature [280].

As far as sexual behaviour is concerned, there is very convincing evidence of the direct action of oestrogen upon cells in the hypothalamus. These hormones induce oestrous behaviour if injected into the hypothalamus of female hamsters *Cricetus* and rats *Rattus* in doses too small to have an effect if injected under the skin [87, 130]. Hence they probably act locally upon the brain cells. Even more convincing is Harris's work on the cat *Felis catus* [106, 107, 108]. He implanted minute fragments of stilboestrol esters (which are oestrogenic) into the posterior hypothalamus. Fragments as small as 0·1 mg induced full mating behaviour in female cats even when the ovaries had been removed. Single implants under the skin had no effect even when as large as 1·0 g, but multiple implants induced changes in the genital tract (increase in size and cornification), followed by oestrous behaviour. This is the normal sequence of events in the intact animal. The brain implants, however, did not influence the genital tract, and this strengthens the likelihood that they were acting directly upon nerve cells concerned in oestrous behaviour. Harris obtained negative responses if he injected into other regions of the brain.

To sum up: gonad hormones appear to be major (and sometimes essential) factors in promoting the responsiveness of the neural systems concerned in courtship and mating. Sometimes, at least, they act directly upon certain regions of the hypothalamus. They may also promote sensory inputs into the central nervous system through their effects upon the genital tract and various sense organs, but these appear to play a minor role in augmenting responsiveness.

An intricate system for synchronizing sexual responsiveness with external and internal events has been described: it involves the hypothalamus, the pituitary, the gonads and the relevant neural systems. Changes in the concentrations of hormones are often slow; hence consequent changes in responsiveness are likely to be gradual, but may be prolonged even when the inducing stimuli are not continuously present. The seasonal build-up of sexual motivation is synchronized with events such as increasing day-length, increasing temperature, the presence of nest material, the presence (and display) of a mate. These events promote a gradual increase of gonad hormones, even though some of them occur only intermittently. at the end of the season their withdrawal leads to a corresponding decline, although internal rhythms in the endocrine organs may play a part in both build-up and decline. Hormonal mechanisms are perhaps better adapted than neural mechanisms to operate in such circumstances.

10: Nervous Mechanisms

This chapter discusses the sexual control mechanism itself: what it consists of and how it operates. First, we consider what and where it is.

The brain and sexual behaviour

The search for regions of the brain especially involved in sexual behaviour has concentrated on female mammals. Much of it has been concerned with the hypothalamus, a region which has already been implicated in the hormone experiments. This comprises the floor of the posterior part of the fore-brain (the diencephalon) (figure 59); it also has visceral functions relating to feeding, water balance (drinking and urination), temperature control and sleep.

Early demonstrations that the hypothalamus is concerned in sexual mechanisms comes from experiments in which parts of the brain were cut away or damaged. Dempsey and Rioch [78] found that spontaneous oestrous behaviour in female guinea-pigs could be abolished if the whole of the fore-brain (diencephalon and telencephalon) was detached from the rest of the brain (by making a transverse cut), but that it remained if only the telencephalon and cortex were removed. Magoun and Bard [18] made large lesions in the hypothalamus and thereby abolished oestrous behaviour in guinea-pigs, cats and even rabbits under hormone treatment.

More delicate techniques suggest the existence of two separate regions, one controlling oestrous behaviour and one ovulation. Unfortunately, lesions may produce general disturbances; hence conclusions based on the consequent behavioural upsets are always suspect. However, Sawyer was able to make very small injuries in the hypothalamus of rabbits; those in or near the mamillary bodies abolished oestrous behaviour but left ovulation inducible (by vaginal stimulation). Conversely lesions near the pituitary stalk blocked ovulation but had no effect upon behaviour [242]. Sawyer also obtained similar results in cats [244], although in slightly different locations.

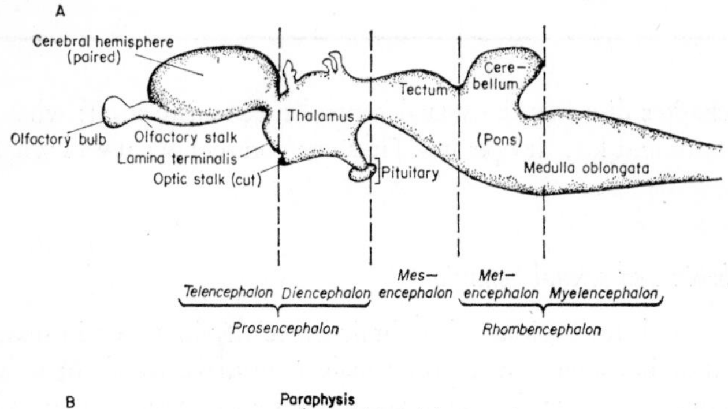

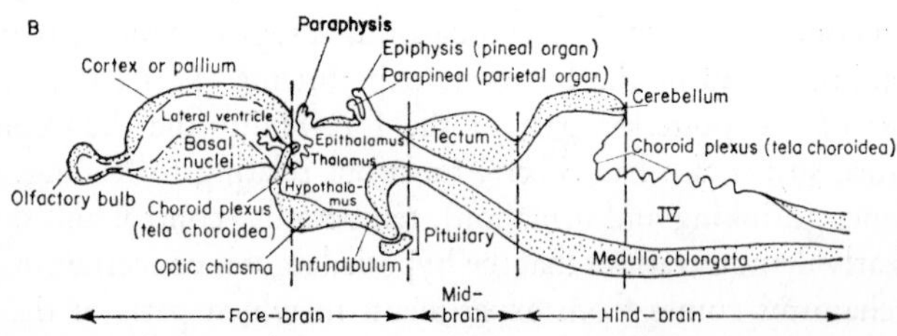

Figure 59. Schematic diagram of a vertebrate brain in external view (A) and median section (B) showing the relative position of the hypothalamus. In mammals the cerebral cortex envelops the brain dorsally while the hind brain bends downwards. (After Romer [231].)

These findings fit well with the work of Harris, already quoted (page 175), in which implants of oestrogenic hormones in the posterior hypothalamus of female cats induced mating behaviour but had no effect upon the reproductive tract. Presumably the 'behaviour centre' alone was stimulated. From this there emerges a coherent picture of sexual control in females. The 'ovarian centre' may be presumed to respond mainly to sensory inputs (long-term ones resulting from seasonal changes, more immediate ones produced by the mate). When activated, it produces neuro-hormones to influence the pituitary and through it the ovary. It thus controls the oestrous cycle and the production of oestrogen by the ovary. This hormone will, in its turn, activate the 'behavioural centre' controlling courtship and mating. And, because oestrogen reaches a peak only when the female is about to ovulate, the behaviour appears at the appropriate moment. There is clearly no need for any direct nervous link between the two centres to ensure this.

There is now some evidence that the ovulation centre can be subdivided into regions with excitory and inhibitory effects, respectively, upon the pituitary [242]. Sometimes local lesions in the anterior hypothalamus have the surprising effect of accelerating oestrus in anoestrous mammals (ferrets) or inducing a state of permanent oestrus (guinea-pigs and rabbits). This suggests that an inhibitory region has been destroyed: it is common for endocrine mechanisms to operate through pairs of antagonistic hormones. This is perhaps partly because a depressant factor (such as a cold spring in the case of ovarian development) can operate more immediately by producing an antagonistic hormone than by causing the withdrawal of the activating one.

Two other, quite different techniques, also indicate that the hypothalamus is especially concerned in mating behaviour. First, electrical recordings were made from the anterior and lateral hypothalamus of cats (under an anaesthetic, and curare to eliminate movement) [214]. A distinctive pattern of electrical activity was obtained when the vagina was stimulated with a glass rod, but only when the cat was in oestrus. Similar results were obtained in rabbits [243].

Secondly, there is the work of Olds [203, 204, 205] who implanted electrodes in various regions of the brains of rats, and put them into a modified Skinner box in which they could, by pressing a lever, electrically stimulate their brains. Some parts of the brain were 'punishing': that is, the rats learned to avoid getting shocks in those regions; but others were surprisingly rewarding; the rats seemed to like the shocks so much that they pressed the lever as often as 3000–5000 times an hour. Moreover, in some areas the attractiveness of the shocks (as measured by the rate of lever pushing) was found to vary with hunger, in others according to the dosage of administered androgens. It is tempting to suppose that the former regions are concerned with feeding behaviour, the latter with sex. Both lie partly in the hypothalamus and partly more anteriorly in the limbic regions of the telencephalon; the 'sex regions' always lie lateral to the 'feeding' ones. These conclusions have been disputed for various reasons [216], but the evidence is of considerable interest, even if not conclusive.

The hypothalamus is not the only site of sexual control. One of the chief characteristics of brain functional geography appears to be the considerable duplication of areas concerned with the same process. This is partly a consequence of evolution. As more and more association areas develop in the fore-part of the brain there is an increasing tendency for more and more mechanisms to be represented there. When this occurs the primitive centres may not be lost, but continue to function under the domination of new ones. Hence, as the thalamus develops as a sensory integration centre the hypothalamus becomes a region of visceral control taking over some of the functions of the medulla. And as the cerebal cortex grows out from the roof of the telencephalon (incorporating huge 'association' networks as well as motor and sensory centres), this region has increasing influence over many of the thalamic and hypothalamic mechanisms. Where sexual behaviour is concerned one might expect therefore both subsidiary centres in lower (more posterior) regions of the brain as well as 'higher' centres in more anterior regions.

A subsidiary region of control has been demonstrated in the mid-brain of female cats [19]. These animals can be made to crouch and tread when in oestrous condition, if the whole of the fore-brain is separated (by a transverse cut) from the mid- and hind-brain. This behaviour, however, lacks the vigour and spontaneity displayed by similar cats retaining the fore-brain (but not the cortex). These, when on heat, crouch persistently, without external stimulation; 'mid-brain' cats require stimulation on the vulva before they will show any oestrous behaviour. In males there is other evidence of 'lower' control centres; erection and ejaculation are under sympathetic control and can be induced in the 'spinal' animal in which the whole of the brain is out of action [31, 296].

Higher centres are, however, much more important in male mammals than in females. Apart from erection and ejaculation, all other male mating behaviour (including courtship) requires the cortex. This does not mean that males lack a hypothalamic sexual control system: the work of Olds (in which both sexes were used) suggests the contrary. But a subordinate hypothalamic system may be dominated by the cortex and this is in line with the fact that male courtship and mating patterns are usually more complex and more modifiable than those of females. Frequently, for instance, males learn details of their mates; they may also adjust their behaviour according to the reactions of a particular female. And the cortex may be involved in co-ordinating the muscular patterns. But precise evidence of cortical influence is patchy and inconsistent. For example, Beach reports that certain areas of the neocortex are concerned in the organization of the coital pattern in male carnivores, yet in rodents these regions are not essential [31].

There is some evidence that the limbic system is concerned in the

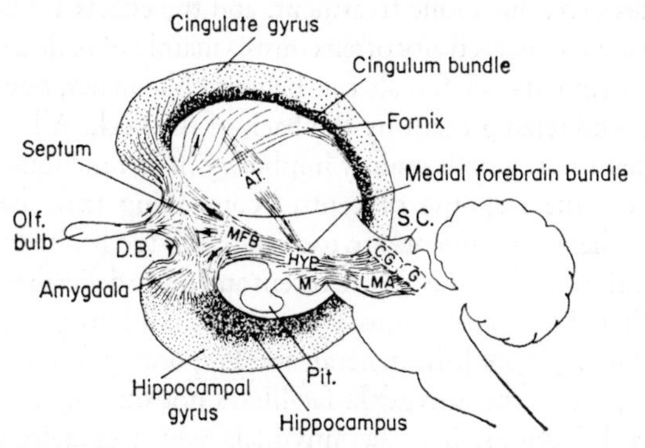

***Figure* 60. Schematic diagram of some of the areas of the primitive cortex which form the limbic system. The overlying brain stem region has been removed to reveal them on the ventro-medial aspect of the right cerebral cortex but the position of the hypothalamus (HYP) and the pituitary (PIT) is shown. The fornix and the median fore-brain bundle (MFB) link this system to the hypothalamus and the mid-brain reticulum. A.T. anterior thalamic nuclei; C.G. central grey of mid-brain; D.B. diagonal band of Broca; G. ventral and dorsal tegmental nucleus of Gudden; L.M.A. limbic mid-brain area of Nauta; M. mammillary body; S.C. superior colliculus.**
(After MacLean [168a], from Gellhorn and Loofbourrow [93].)

mating behaviour of both male and female mammals. The limbic system (figure 60) is the name given to the older, more primitive areas of the most anterior part of the fore-brain (telencephelon) which are intimately related to the hypothalamus, mid-brain reticulum and autonomic nervous system. They lie on the medial and ventral aspects of the telencephalon and have been overgrown by the roof-pouches forming the neocortex. They were primitively concerned with motor integration (basal ganglia) and with smell (rhinencephalon). But in the higher mammals they have acquired new functions and new names (cingulate gyrus, hippocampus, amygdala, septal, orbito-frontal and piriform regions of the cortex) [82, 120].

Most experiments in which areas of the limbic system are removed or destroyed increase sexual activity especially in animals dependent upon gonad hormones [98]. Lesions often delay the loss of sexual activity after castration; they also promote hypersexual behaviour in castrates given hormone treatment, and the effects are persistent. The increased sexual activity often consists mainly of male behaviour; in one experiment even female cats behaved like males, pouncing on other cats and seizing them by the back of the neck. All this might suggest that this region is mainly inhibitory in its functions, and that it influences the response of centres controlling male behaviour. However, these experiments are too crude and too limited to permit generalizations such as this. More recent and more delicate techniques suggest that there are excitatory as well as inhibitory areas in the limbic system [82, 98]. In general, the hippocampus appears to be inhibitory, while the amygdala facilitates not only sexual but also aggressive behaviour. It is the amygdala which contains many of Olds' 'pleasure centres' and, besides its effect upon sexual behaviour, it influences general responsiveness. Animals whose amygdalae have been removed are often abnormally placid. Other work suggests that the septal and cingular regions are also concerned in reproductive behaviour [82]. Beach has criticized much of the work on the limbic system, saying that the experiments are not sufficiently controlled and that much of the reported hyper- and hypo-sexual activity comes within the range of normal behavioural variation [31]. However, evidence is accumulating. If it is involved, the limbic region, like the cortex, may exert excitatory or inhibitory pressures on the control centres as a result of experience. Both cortex and limbic system appear to be concerned in processes of memory and learning [82].

The evidence suggests, then, the existence of a sexual control centre in the hypothalamus, a subsidiary centre in the mid-brain and 'higher' centres capable of raising or depressing excitability in the limbic system and the cortex. The evidence is by no means conclusive, and other regions of the brain are probably also involved.

Responsiveness

It is now necessary to enquire how this system works. A principal foundation of our knowledge of nervous function comes from Sherrington's work on reflexes [69, 249]. Reflexes are immediate responses to special stimuli, for example, the abrupt withdrawal of a limb from a painful stimulus. Their control systems are relatively simple, and we now understand much about their mode of operation. How does more complex behaviour differ from reflexes?

We first consider changes in responsiveness. Reflexes are remarkable for their stability in this respect. It is possible (after many hundred repetitions in quick succession) to diminish the elicitability of a reflex or even extinguish it, and some reflexes are extinguished during sleep or unconsciousness. But these variations in no way compare with the fluctuations in responsiveness typical of more complex behaviour patterns. The latter are influenced by fluctuating chemical factors such as sex hormones, whereas there is no evidence of this for reflexes.

The stability of reflex systems is, however, by no means automatic. Even the spinal reflexes of vertebrates depend upon neural factors fed from the brain. For example, continuous discharges from certain regions of the ear of cats and dogs (the maculae and semicircular canals) are necessary for the operation of the stretch reflexes which enable the animal to stand upright. An animal whose spinal cord has been severed from the brain cannot stand. These excitatory inputs, transmitted via lower regions of the brain, are held in balance by counteracting inhibitory ones from higher regions, for an animal with only the lower regions intact maintains an abnormally rigid posture. However, the balance of a normal animal is adjusted to an intermediate level, so that responsiveness is high but not excessive.

Undoubtedly neural factors modify the responsiveness of more complex behaviour patterns in a similar way. A very generalized influence comes from the reticular formation. This is a system of diffusely interconnecting neurons centrally placed in the brain stem and connecting with the thalamus, hypothalamus, mid-brain, limbic system and cortex. It receives many sensory inputs but in a random fashion, not ordered as in the more external regions of the brain. It is now believed that the reticular formation differs from the more external brain regions, not only in anatomy, but also in function, being less concerned with immediate reactions to external stimuli and more with responsiveness in general. When it is destroyed, the animal falls into a deep sleep or coma; when it is stimulated, the animal becomes alert, vigilant and quick to react. This result is achieved in several ways: by affecting the excitability of the cortex, by influencing muscle tone, sympathetic tone and possibly the sensitivity of receptors too. The system is primed unspecifically by sensory inputs and also by many of the factors known to influence specific behaviour patterns: adrenaline, carbon dioxide, oestrogen and progesterone. In other words, the system makes the animal more ready to react whenever its input of sensory and chemical factors reaches a certain volume, preparing it for activity in general but not for any particular activity [44, 82, 129].

There are also more specific neural modifiers of sexual responsiveness. The evidence already discussed suggests that these come from the limbic system and the cortex; they may fluctuate for various reasons. It is possible that sex hormones may act upon these regions of the brain as well as upon the specific control mechanisms. More important perhaps are the many stimuli which can affect sexual responsiveness. There are a number of these which have 'arousal' effects; they appear to induce restless activity in a male: a cockroach, for example, on perceiving the scent of females, rushes around tapping objects with its antennae. In other species such stimuli may induce more directed searches, but they do not trigger off courtship or mating itself. Arousing stimuli include most of the 'distant' stimuli from females, namely, scents, sounds and vibrations (contrasted with close-range sights or contacts which do commonly operate as triggers). There may also be inputs from genital organs or other regions sensitized by hormones. Other stimuli depress sexual activity: sexual responsiveness is reduced after mating, for example, by fresh eggs (sticklebacks) or by stimuli from genital organs.

Probably some of these stimuli act on higher regions of the brain. For, particularly in higher animals, their effectiveness can vary as a result of experience; and new stimuli, occurring in the right context, may come to have significance in arousing or depressing responsiveness. Indeed, it is during the 'searching' phase of an activity that an animal most readily learns. A hungry rat will learn to run a maze to get food, a ring dove with distended crop learns to react to squabs, and a sexually motivated male bird or mammal may learn details of his mate's appearance. All this suggests the operation of the cortex and the limbic system.

It seems, then, that factors affecting responsiveness may have many sites of action: they may operate directly upon the control system concerned, they may feed to higher centres and they operate a general alerting system. In addition they may also bias sense organs and sensory integration centres to give priority to relevant stimuli. This possibility will be discussed shortly; it ensures that the animal although generally alerted, often tends to ignore some stimuli and respond to others.

Switching-on

Given a responsive control system, how is it switched on? In a reflex system, activity is always a response to a specific stimulus. The link-up is achieved through fairly direct nervous pathways between the sensory cells and the effector organs. Often sexual responses, too, occur to specific stimuli, although these may be complex and involve several sense modalities. But the response itself is rarely simple, and the sequence of component activities may vary continuously. And, while courtship movements vary according to the female's reactions, there is usually no one-to-one relationship between signal and response: instead the behaviour of the male continually readjusts itself to the changing input. Clearly, no simple, fixed system of nervous connections can be responsible for this.

A few insects possess sense organs which appear to be solely concerned in sexual behaviour reactions. Chemoreceptors in the antennae of male silkworms *Bombyx mori* respond only to the specific secretion produced by females of the same species [57, 246]. Male mosquitoes of the species *Aedes aegypti* possess sensory hairs which appear to resonate specifically to the frequency of the female's wing beat [234].

Usually, however, a sense organ receives stimuli relevant to several behaviour patterns. The fritillary butterfly responds to yellow, blue and green coloured objects when feeding but only to yellow-orange ones when seeking a mate [170]. All these objects are perceived by the eyes, hence some sorting must occur before appropriate inputs can be connected with appropriate motor centres. This process is made simpler by filtering, so that only a sample of the most important types of input reach the association areas. Indeed filtering is essential if these sorting systems are not to be overloaded. Calculations of the number of cells available for sorting and discriminating in the cortex of a mammal (less than 10^{10}), compared with the number of possible combinations of activity in the three million or so sensory fibres which supply it (about 2^3 million), suggest that the cortex cannot possibly deal with all the information which the sense organs could supply [21].

Much compression and coding also occurs in the sense organs themselves. These pass only a fraction of the information which reaches them to the nervous system, usually that which is most relevant to behavioural releasers. For example, in the compound eye of arthropods, the quick adaptation of individual ommatidia limits the amount of information sent [288], and the releasers which incorporate broken outlines and flickering movements are adapted to transmit this particular filtering system most effectively [7, 170]. In the vertebrate eye, too, there is quick adaptation of individual retinal cells and also a process called lateral inhibition, whereby the discharge of any given cell tends to be reduced if a neighbouring cell is similarly illuminated. The former facilitates the perception of moving objects, the latter that of edges (as opposed to uniform surfaces) [20, 21, 138].

It is usually assumed (and probably often true) that releasers have evolved to fit in with the perceptive abilities and coding devices of the sense organs. But sometimes a coding device seems to develop only when it is required. For example, lateral inhibition does not occur in cats with completely dark-adapted eyes where maximum sensitivity is desirable: it is acquired by cats in the light where the perception of edges is valuable [22]. It could be that other coding devices are also acquired to aid quick recognition of significant shapes, sounds and movements. Hence, sometimes the sense organs may adapt themselves in early life to fit the releaser. Filtering and coding devices, then, allow for increased complexity in triggering systems within the limitations of the available sense organs and nerve cells [177].

Switching-off

A reflex response depends mainly on its triggering stimulus: it occurs when it is present and it stops when it is absent. This is obviously true of courtship too; it usually stops when the female (or stimuli from the female) disappears. But courtship can be stopped in many more ways than this. The removal of chemical factors like hormones can do so. Further, special stimuli can stop courtship temporarily and their effect is often immediate. Courtship may also stop after prolonged and unsuccessful performance. What factors are operating here?

Special stimuli which stop activities are often called inhibitory stimuli, implying that they provide inhibitory inputs to the centres concerned. Reflex systems can inhibit one another, but there is no evidence of external stimuli which do so directly. In more complex systems, too, inhibitory stimuli may not operate directly but through the mediation of higher regions of the brain. Some special stimuli, like those received by the rabbit during coitus, influence hormone secretion (which, in this case, promotes ovulation). But most inhibitory stimuli which stop courtship do so too quickly for this explanation to be likely: when a fertilized female fruit-fly extrudes a courting male turns away from her at once. However, sometimes inhibitory effects appear to outlast the stimuli: female fruit-flies, for instance, lose their receptivity for up to twenty-four hours after copulation even if no sperm is received. It seems unlikely that the effects of vaginal stimuli could persist so long; in this case the mechanism is unknown.

The stopping of an activity after prolonged performance is more difficult to explain. Granted, nervous mechanisms can be 'fatigued'. By repeatedly evoking certain reflex responses in a brainless frog, Franzisket [89] reduced responsiveness without sensory adaptation or muscular fatigue. But many hundreds of repetitions were required for this to happen: courtship usually stops before such an explanation could apply.

Perhaps a special type of 'adaptation' to repeated stimuli occurs. There is no known mechanism for this, but it would help to explain certain observations on rats by Kimble and Kendall [131]. They repeatedly elicited a certain response by animals of one group until they failed to respond any more; another group were repeatedly stimulated in the same way, but too weakly to evoke a response. The second group, tested at intervals with a strong stimulus, was found to lose its responsiveness even more quickly than the first. Evidently through repeated presentation, the stimulus had lost its potency. The same perhaps applies to cases where a male ceases to court a given female, but immediately renews display on encountering a fresh one. Male rats, enclosed with one or several females until they cease to respond to them sexually, will show renewed mating activity if a new 'unused' female is introduced [43]. Since a new 'used' female does not have this effect, it is possible that a generalized stimulus from 'used' females (chemical or behavioural) has lost its potency. However fruit-flies which have ceased to respond to a female (whether or not she has accepted him) will renew courtship with any new one, fertilized or virgin [personal observations].

Interference

Sometimes, several behaviour patterns are activated at once. There is behavioural evidence that they can interfere with one another. Is there any neurological evidence?

It has already been described how, at the reflex level, the stimulation of one activity inhibits an opposing one. Evidence regarding more complex behaviour comes from the work of von Holst and St Paul [118]. They implanted fine electrodes (several at a time) in the brain stem of a fowl. On electrical stimulation these often evoked distinct elements of behaviour which were recognizably parts of

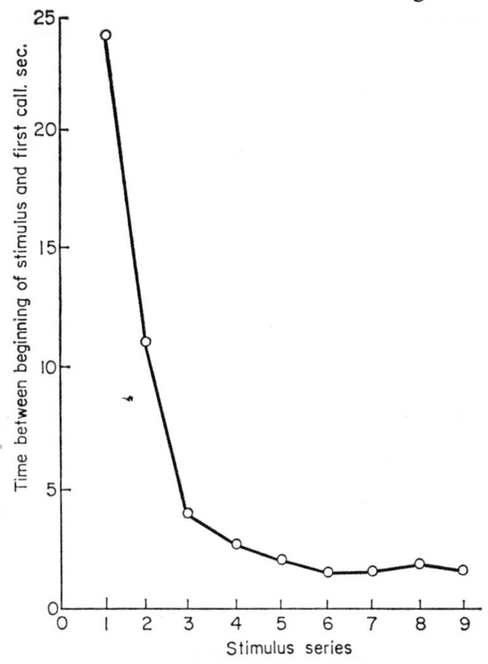

Figure 61. Changes in threshold. A quietly sitting fowl was repeatedly stimulated with a constant stimulus (0.4 volt) in an area of the brain which induced clucking, each time until the first cluck. Between one cluck and the next stimulus was a pause of 5 seconds. The duration of the required stimulus fell rapidly.
(After von Holst and St Paul [118].)

normal behaviour patterns. At first a strong stimulus was required to elicit any one of these, but with repeated or protracted stimulation the required strength fell progressively (figure 61). Simultaneously that needed at other electrodes tended to rise, particularly if the behavioural elements concerned were antagonistic to the first. Thus a decline in the stimulus strength required to elicit feeding was accompanied by a rise in that required for sleep; and a decline for sleep by a rise for flight. The activation of one mechanism evidently reduced the responsiveness of others. They also found that the strong activation of a new response during the performance of another could result in the new activity interrupting or even suppressing the first. This occurred when two different activities, watching out and sitting (brooding), were stimulated in a hen which was cackling in response to a frightening stimulus. They found that the latter was more effective than the former in interfering with cackling and a prolonged bout of sitting could suppress cackling altogether (figure 62).

Further evidence comes from the work of Olds [204] according to which treatment with sex hormones not only increased self-stimulating responses (bar pressing) in rats with electrodes in 'sex

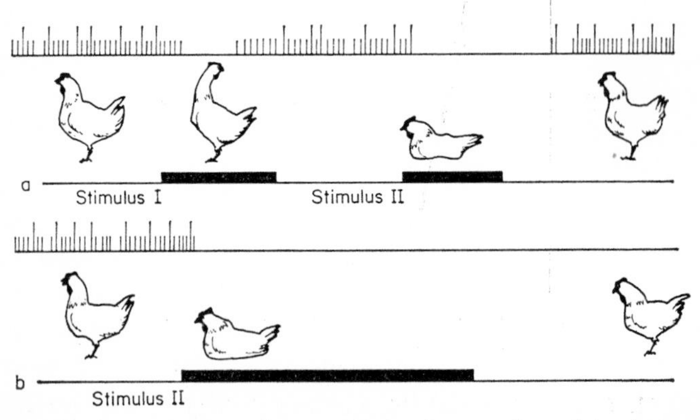

Figure 62. Interruption of one activity by another. Stimulation of 'watching out' (stimulus I) and 'sitting' (stimulus II) in a fowl which is cackling continuously. Cackling indicated by upper graph; other reactions by pictures (see text).
(After von Holst and St Paul [118].)

centres' but also decreased such responses in 'feeding centre' rats. Conversely, hunger enhanced responses from 'feeding centre' rats and diminished them in 'sex centre' ones. These findings suggest that the activation of one system interfered with the operation of another concerned with a different type of behaviour.

All this suggests that interference between systems is a matter of inhibitory influences impinging upon the system itself. However, there is some evidence of interference with *inputs* into competing systems. Most people would say, from introspection, that when they are hungry, for example, they are more acutely aware of stimuli relevant to food than to any others (even to the extent of wishful misinterpretation). A jumping spider after a long fast may treat a female of his own species as prey (which she resembles in some ways). And when von Holst strongly stimulated electrodes evoking aggressive behaviour, a hen would attack a keeper normally treated as a friend. Activating factors for any particular behaviour pattern may bias sense organs or sensory centres. This does not, of course, necessarily mean that irrelevant stimuli are filtered out, but there is at least one investigation suggesting that this can occur. Hernandez-Peon, Scherrer and Jouvet [111] recorded regular impulses from the cochlear nerve of a cat whenever clicking sounds were played to it. However, if they showed the cat a mouse, the impulses ceased, even although the clicks continued. In this case it appears that irrelevant stimuli were blocked at the sense organs and did not even enter the central nervous system. Mechanisms such as these may account for specific sensory reactions in searching behaviour discussed earlier.

If behavioural systems can inhibit one another, what happens when *opposing* systems are simultaneously stimulated? Can conflict behaviour be understood in physiological terms? Sherrington has some experimental evidence on this point [249]. He applied pain stimuli to opposite limbs of a brainless mammal simultaneously. Pain stimuli cause a limb to flex and flexion of one limb inhibits flexion in the opposite one. Hence Sherrington was simultaneously stimulating opposing systems. He found that if the stimuli were unequal, the stronger dominated; if they were equal, both limbs flexed. Thus each overcame the inhibition imposed by the other. Sherrington then tried weaker, continuous stimuli (pain stimuli are relatively strong) applied to the sensory nerves of both hind limbs. Then he obtained rhythmic stepping movements. In this case he supposed that each dominated in turn, the other being meanwhile inhibited. This is perhaps analogous to ambivalent behaviour.

Where more complex behaviour systems are concerned, the work of von Holst and St Paul [118], with their electrodes inserted in hens' brains, is again of interest. They chose two electrodes, each

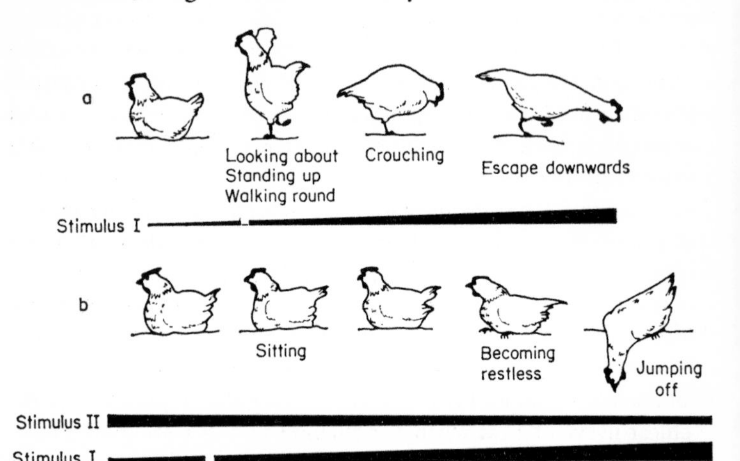

Figure 63. Interference between two activities in a fowl.
(a) A normal sequence of fleeing behaviour (from an aerial enemy) induced by electrically stimulating a given area of the brain with a slowly increasing stimulus; (*b*) the result when 'sitting down' was simultaneously stimulated.
(After von Holst and St Paul [118].)

eliciting a different activity, and stimulated them simultaneously Their results were very varied. Figure 63 illustrates one example where one of the activities (fleeing) gradually superseded the other (sitting) as it was increasingly stimulated, but its expression was considerably modified. Other results depended upon the activities chosen. If the two activities could be performed together (like sitting and preening), then this would be done. If they could be performed more or less together, at the expense of some incompleteness, then this would be done too. An example is reconnoitring (peering with stretched neck) and searching (sweeping the head to and fro). In simultaneous performance the head was not stretched so far and the head sweeps were shorter. With less compatible activities, sometimes one was followed by the other (sitting and fleeing), or there was an alternation of the two activities (eating and reconnoitring). And in one case (pecking and fleeing) neither occurred, but instead the bird rushed backwards and forwards, calling, with raised feathers.

The types of behaviour obtained by these techniques do resemble some forms of conflict activity, and they suggest that different control systems can and do interact when stimulated together. A systematic investigation of the conditions under which various types of interaction occur might tell us a good deal more.

Conclusion

It seems possible to postulate a sexual control system which operates upon the principles understood from reflex mechanisms. But the system is neither discrete nor uniform in operation. Many areas of the brain may be involved, and activating (or inhibiting) factors may operate at several different sites simultaneously. Some of these may be concerned with the specific activity, others with more generalized reactions. Moreover, the operation of the system may be modified by interference from other control systems or as a result of experience. Nevertheless a picture emerges of a mechanism which can operate according to the rules which emerge from analyses of sexual behaviour.

Where courtship is concerned, von Holst's work suggests that modified and even quite distinct activities may result from the activation of two systems simultaneously. Courtship may then result from the simultaneous activation of both the sexual mechanism and also another. However, where the display is ritualized there must be devices to regularize both the link-up and the reaction. We know very little about such processes [38]. Where movements are exaggerated or reduced or repeated or changed in frequency, changes in nervous or neuromuscular thresholds are likely to be involved: sometimes there may be anatomical changes, in effector organs;

sometimes hormonal ones. Many of the modifications may be such as to effect many behaviour patterns, not just courtship alone. There may be a general increase in the scale or frequency of one type of movement and its exaggeration in courtship merely reflects this. Thus while ritualization may involve the establishment of a separate courtship control system it is equally possible that courtship is still the direct outcome of 'conflict', its form and regularity being modified by peripheral or generalized changes affecting the interacting systems.

It seems that while we know enough to postulate a reasonably coherent picture of the physiology of sexual behaviour in vertebrates we still have much to learn about courtship. Its significance and evolutionary origins are reasonably clear but details of its mechanisms are largely unknown. Probably the relationship between it and mating behaviour differs from animal to animal and the processes involved must be studied separately in each species.

Bibliography

1 ANDREW, R. J., 1956. 'Some remarks on behaviour in conflict situations, with special reference to *Emberiza* species': *Br. J. Anim. Behav.* 4, 41–5.

2 ANDREW, R. J., 1957. 'The aggressive and courtship behaviour of certain Emberizinae': *Behaviour* 10, 255–308.

3 ARMSTRONG, E. A., 1947. *Bird Display and Behaviour:* Lindsay Drummond, London.

4 ARONSON, L. R., 1944. 'The influence of the male *Tilapia macrocephala* upon the spawning frequency of the female': *Anat. Rec.* 89, 539.

5 ASDELL, S. A., 1946. *Patterns of Mammalian Reproduction:* New York.

6 BAAKE, K., 1928. 'Die Brachydanio-Arten als Laichräuber': *Bl. Aquar.–u. Terrarienk.* 39, 307–11.

7 BAERENDS, G. P., 1950. 'Specializations in organs and movements with a releasing function': *Symp. Soc. exp. Biol.* 4, 337–60.

8 BAERENDS, G. P., 1952. 'Les sociétés et les familles de poissons': *Colloqués Intern. d. Centre Nat. de la Recherche Scient.* 34, 207–19.

9 BAERENDS, G. P., and BAERENDS-VAN ROON, J. M., 1950. 'An introduction to the study of the ethology of cichlid fishes': *Behaviour Suppl.* 1, 1–242.

10 BAERENDS, G. P., BROUWER, R., and WATERBOLK, H. T. J., 1955. 'Ethological studies of *Lebistes reticulatus* (Peters). I. An analysis of the male courtship pattern': *Behaviour* 8, 249–334.

11 BAGGERMAN, B., BAERENDS, G. P., HEIKENS, H. S., and MOOK, J. H., 1956. 'Observations on the behaviour of the black tern *Chlidonias n. niger* (L), in the breeding area': *Ardea* 44, 1–71.

12 BAKER, J. R., 1938. 'The evolution of breeding seasons'. In *Evolution. Essays on Aspects of Evolutionary Biology:* G. D. de Beer, ed., Clarendon Press, Oxford.

13 BAKER, J. R., and BIRD, T. F., 1936. 'The seasons in a tropical rain forest (New Hebrides).' 4. 'Insectivorous bats (*Vespertilionidae* and *Rhinolophidae*).': *J. Linn. Soc. Lond. Zool.* 40, 143–61.

14 BALL, J., 1934. 'Sex behavior of the rat after removal of the uterus and vagina': *J. comp. Psychol.* 18, 419–22.

15 BANE, A., 1954. 'Studies on monozygous cattle twins. Sexual functions of bulls in relation to heredity, rearing intensity and somatic conditions': *Acta agric. Scand.* 4, 95–208.
16 BARBER, H. S., 1951. 'North American fireflies of the genus *Photuris*': *Smiths. Misc. Coll.* 117 (1), 1–58.
17 BARD, P., 1935. 'The effects of denervation of the genitalia on the oestrual behavior of cats': *Am. J. Physiol.* 113, 5–6.
18 BARD, P., 1940. 'The hypothalamus and sexual behavior': *Res. Publ. Ass. nerv. ment. Dis.* 20, 551.
19 BARD, P., and MACHT, M. B., 1958. 'The behavior of chronically decerebrate cats': in *Ciba Foundation Symposium on the Neurological Basis of Behavior:* Little, Brown & Co., Boston, 55–75.
20 BARLOW, H. B., 1953. 'Summation and inhibition in the frog's retina': *J. Physiol.* 119, 69–88.
21 BARLOW, H. B., 1961. 'The coding of sensory messages': in *Current Problems in Animal Behaviour:* W. H. Thorpe and O. L. Zangwill, eds., Camb. Univ. Press.
22 BARLOW, H. B., FITZHUGH, R., and KUFFLER, S. W., 1957. 'Change of organization in the receptive fields of the cat's retina during dark adaptation': *J. Physiol.* 137, 338–54.
23 BARTHOLOMEW, G. A. J., 1949. 'The effect of light intensity and day length on reproduction in the English sparrow': *Bull. Mus. Comp. Zool.* 1, 433–76.
24 BASTOCK, M., 1956. 'A gene mutation which changes a behavior pattern': *Evolution* 10, 421–39.
25 BASTOCK, M., and MANNING, A., 1955. 'The courtship of *Drosophila melanogaster*': *Behaviour* 8, 85–111.
26 BASTOCK, M., MORRIS, D., and MOYNIHAN, M., 1953. 'Some comments on conflict and thwarting in animals': *Behaviour* 6, 66–84.
27 BATEMAN, A. J., 1948. 'Intra-sexual selection in *Drosophila*': *Heredity* 2, 349–68.
28 BEACH, F. A., 1944. 'Relative effects of androgen upon the mating behavior of male rats subjected to forebrain injury or castration': *J. exp. Zool.* 97, 249–95.
29 BEACH, F. A., 1947. 'A review of physiological and psychological studies of sexual behavior in mammals': *Physiol. Rev.* 27, 240–307.
30 BEACH, F. A., 1948. *Hormones and Behavior:* Paul B. Hoeber, Inc., New York.
31 BEACH, F. A., 1964. 'Biological bases for reproductive behavior': in *Social Behavior and Organization in Vertebrates,* W. Etkin, ed., Univ. of Chicago Press, 117–42.
32 BERNDT, W., 1925. 'Vererbungsstudien an Goldfischrassen': *Z. indukt. Abstamm-u. Vererblehre* 36, 161–349.
33 BEVAN, W., LEVY, G. W., WHITEHOUSE, J. M., and BEVAN, J.

M., 1957. 'Spontaneous aggressiveness in two strains of mice castrated and treated with one of three androgens': *Physiol. Zool.* 30, 341–9.

34 BLACK, V. S., 1951. 'Osmotic regulation in Teleost fishes': *Publ. Ontario Fish Res. Lab.* 71, 53–90.

35 BLAIR, W. F., 1955. 'Mating call and stage of speciation in the *Microhyla olivacea–M. carolinensis* complex': *Evolution* 9, 469–80.

36 BLAIR, W. F., 1956. 'Call difference as an isolating mechanism in south-western toads (genus *Bufo*)': *Texas J. Sci.* 8, 87–106.

37 BLAIR, W. F., 1958. 'Mating call in the speciation of anuran amphibians': *Am. Nat.* 92, 27–51.

38 BLEST, A. D., 1961. 'The concept of ritualization': in *Current Problems in Animal Behaviour.* W. H. Thorpe and O. L. Zangwill, eds., Camb. Univ. Press, 102–24.

39 BOL, A. C. A., 1959. 'A consummatory situation. The effect of eggs on the sexual behaviour of the male three-spined stickleback (*Gasterosteus aculeatus* L.)': *Experientia* 15, 115.

40 BOYCOTT, B. B., 1953. 'The chromatophore system in Cephalopoda': *Proc. Linn. Soc. Lond.* 164, 235–40.

41 BRAUN, F., 1915. 'Ueber die Streitlust gefangener Sperlingsvögel und ihre Grunde': *Orn. Monatsber.* 23, 33–9, 49–55.

42 BRISTOWE, W. S., 1958. *The World of Spiders:* Collins New Naturalist, No. 38, London.

43 BROADBENT, D. E., 1961. *Behaviour:* Eyre & Spottiswoode, London.

44 BRODAL, A., 1957. *The Reticular Formation of the Brain Stem:* London.

45 BROOKS, C. M., 1937. 'The role of the cerebral cortex and of various sense organs in the execution of mating activity in the rabbit': *Am. J. Physiol.* 120, 544–53.

46 BROWER, L. P., BROWER, J. VAN Z., and CRANSTON, F. P., 1965. 'Courtship behavior of the queen butterfly, *Danaus gilippus berenica* (Cramer)': *Zoologica* 50, 1–39.

47 BROWN, R. G. B., 1964. 'Courtship behaviour in the *Drosophila obscura* group. I: *D. pseudoobscura*': *Behaviour* 23, 61–106.

48 BROWN, R. G. B., 1965. 'Courtship behaviour in the *Drosophila obscura* group. Part II: Comparative studies': *Behaviour* 25, 281–323.

49 BRUCE, H. M., 1959. 'An exteroceptive block of pregnancy in the mouse': *Nature, Lond.* 184, 105.

50 BRUCE, H. M., 1962. 'The importance of the environment in the establishment of pregnancy in the mouse': *Anim. Behav.* 10, 399.

51 BUCK, J. B., 1937. 'Studies on the firefly. II. The signal system and color vision in *Photinus pyrelis*': *Physiol. Zool.* 10, 412–19.

52 BULLOUGH, W. S., 1942. 'The reproductive cycles of the British and Continental races of the starling': *Phil. Trans. R. Soc. Lond.* B 231, 165–246.

53 BULLOUGH, W. S., 1945. 'Endocrinological aspects of bird behaviour': *Biol. Rev.* 20, 89–99.

54 BULLOUGH, W. S., 1951. *Vertebrate Sexual Cycles:* Methuen & Co. Ltd, London.

55 BURGER, J. W., 1942. 'The influence of some external factors on the ovarian cycle of the female starling': *Anat. Rec.* 84, 518.

56 BURGER, J. W., 1949. 'A review of experimental investigations on seasonal reproduction in birds': *Wilson Bull.* 61, 211–30.

57 BUTENANDT, A., 1955. 'Über Wirkstoffe des Insektenreiches. II. Zur Kenntnis der Sexual-Lockstoffe': *Naturw. Rdsch.* 12, 457–64.

58 CARPENTER, C. R., 1958. 'Territoriality: A review of concepts and problems': in *Behavior and Evolution:* A. Roe and G. G. Simpson, eds., Yale Univ. Press.

59 CASPARI, E., 1958. 'Genetic basis of behavior': in *Behavior and Evolution:* A. Roe and G. G. Simpson, eds., Yale Univ. Press.

60 CLARK, E., ARONSON, L. R., and GORDON, M., 1954. 'Mating patterns in two sympatric species of Xiphophorin fishes: their inheritance and significance in sexual isolation': *Bull. Am. Mus. nat. Hist. N.Y.* 103, 135–226.

61 COOPER, J. B., 1942. 'An exploratory study on African lions': *Comp. Psychol. Monogr.* 17, 1–48.

62 CORNER, G. W., 1942. *The Hormones in Human Reproduction:* Princeton Univ. Press.

63 CRAIG, W., 1908. 'Oviposition induced by the male in pigeons': *J. Morphol.* 22, 299–305.

64 CRANE, J., 1941. 'Crabs of the genus *Uca* from the west coast of Central America': *Zoologica* 26, 145–208.

65 CRANE, J., 1948. 'Comparative biology of salticid spiders at Rancho Grande, Venezuela. Part I. Systematics and life histories in *Corythelia*': *Zoologica* 33, 1–38.

66 CRANE, J., 1949. 'Comparative biology of salticid spiders at Rancho Grande, Venezuela. Part III. Systematics and behavior in representative new species': *Zoologica* 34, 31–52.

67 CRANE, J., 1949. 'Comparative biology of salticid spiders at Rancho Grande, Venezuela. Part IV. An analysis of display': *Zoologica* 34, 159–214.

68 CRANE, J., 1957. 'Basic patterns of display in fiddler crabs (Ocypodidae, genus *Uca*)': *Zoologica* 42, 69–82.

69 CREED, R. S., DENNY-BROWN, F., ECCLES, J. C., LIDDELL, E. G. T., and SHERRINGTON, C. S., 1932. *Reflex Activity of the Spinal Cord:* London.

70 CROOK, J. H., 1963. 'Comparative studies on the reproductive behaviour of two closely related weaver bird species (*Ploceus cucullatus* and *Ploceus nigerrimus*) and their races': *Behaviour* 21, 177–232.

71 CROSSLEY, S., 1963. 'An experimental study of sexual isolation within a species of *Drosophila*': D. Phil. Thesis, Univ. of Oxford.
72 CULLEN, E., 1957. 'Adaptations in the kittiwake to cliff-nesting': *Ibis* 99, 275–302.
73 CULLEN, J. M., 1960. 'The aerial display of the arctic tern and other species': *Ardea* 48, 1–37.
74 DAANJE, A., 1950. 'On locomotory movements in birds and the intention movements derived from them': *Behaviour* 3, 48–98.
75 DARLING, F. F., 1937. *A Herd of Red Deer:* Oxford Univ. Press.
76 DARLING, F. F., 1938. *Bird Flocks and the Breeding Cycle: A Contribution to the Study of Avian Sociality*: Cambridge Univ. Press.
77 DARWIN, C. R., 1890. *The Descent of Man:* 2nd ed., Murray, London.
78 DEMPSEY, E. W., and RIOCH, D. McK., 1939. 'The localization in the brain stem of the oestrous responses of the female guinea pig': *J. Neurophysiol.* 2, 9–18.
79 DEWAR, D., 1908. *Birds of the Plains:* London and New York.
80 DILGER, W. C., 1962. 'The behavior of lovebirds': *Scient. Am.* 206 (1), 88–98.
81 DREES, O., 1952. 'Untersuchungen über die angeborenen Verhaltensweisen bei Springspinnen (Salticidae)': *Z. Tierpsychol.* 9, 169–207.
82 ETKIN, W., 1964. 'Neuroendocrine correlation in vertebrates': in *Social Behavior and Organization in Vertebrates.* W. Etkin, ed., Univ. of Chicago Press, 35–52.
83 EWING, A., 1961. 'Body size and courtship behaviour in *Drosophila melanogaster*': *Anim. Behav.* 9, 93–9.
84 FABRICIUS, E., and GUSTAFSON, K-J., 1954. 'Further observations on the spawning behaviour of the char, *Salmo alpinus* L.': Institute of Fresh-Water Research, Drottningholm, Report No. 35, 58–104.
85 FABRICIUS, E., and LINDROTH, A., 1954. 'Experimental observations on the spawning of whitefish, *Coregonus lavaretus* L., in the stream aquarium of the Holle laboratory at River Indälsalven': Institute of Fresh-Water Research, Drottningholm, Report No. 35, 105–12.
86 FALCONER, D. S., 1960. *Introduction to Quantitative Genetics:* Oliver & Boyd, Edinburgh and London.
87 FISHER, A. E., 1956. 'Maternal and sexual behavior induced by intracranial chemical stimulation': *Science, N.Y.* 124, 228–9.
88 FOWLER, H., and WHALEN, R. E., 1961. 'Variation in incentive stimulus and sexual behavior in the male rat:' *J. comp. physiol. Psychol.* 54, 68–71.
89 FRANZISKET, L., 1953. 'Untersuchungen zur Spezifität und Kumulietung der Erregungsfähigkeit und zur Wirkung einer Ermüdung in der Afferenz bei Wischbewegung des Ruckenmarkfrosches': *Z. vergl. Physiol.* 34, 525–38.
90 FREDERICSON, E., STORY, A. W., GURNEY, N. L., and BUTTER-

WORTH, K., 1955. 'The relationship between heredity, sex and aggression in two inbred mouse strains': *J. genet. Psychol.* 87, 121–30.

91 FULLER, J. L., and THOMPSON, W. R., 1960. *Behavior Genetics:* John Wiley, New York.

92 GEER, B. W., and GREEN, M. M., 1962. 'Genotype, phenotype and mating behavior of *Drosophila melanogaster*': *Am. Nat.* 96, 175–81.

93 GELLHORN, E., and LOOFBOURROW, G. N., 1963. *Emotions and Emotional Disorders:* Hoeber Medical Division, Harper & Row, New York.

94 GOLDSTEIN, A. C., 1957. 'Control of sex behavior in animals': in *Hormones, Brain Function and Behavior:* M. B. Hoagland, ed., N.Y. Acad. Press Inc., 99–123.

95 GOODWIN, D., 1953. 'Observations on voice and behaviour of the red-legged partridge *Alectoris rufa*': *Ibis* 95, 581–614.

96 GOY, R. W., and JAKWAY, J. S., 1959. 'Inheritance of patterns of mating behaviour in the female guinea pig': *Anim. Behav.* 7, 142–9.

97 GOY, R. W., and YOUNG, W. C., 1957. 'Somatic basis of sexual behavior patterns in guinea pigs': *Psychosom. Med.* 19, 144–51.

98 GREEN, J. D., 1958. 'The rhinencephalon and behavior': in *CIBA Foundation Symposium on the Neurological Basis of Behavior:* Little, Brown & Co., Boston, 222–35.

99 GREEN, J. D., and HARRIS, G. W., 1947. 'The neurovascular link between the neurohypophysis and the adenohypophysis': *J. Endocr.* 5, 136–48.

100 GRUNT, J. A., and YOUNG, W. C., 1952. 'Differential reactivity of individuals and the response of the male guinea pig to testosterone propionate': *Endocrinology* 51, 237–48.

101 HALDANE, J. B. S., 1946. 'Interaction of nature and nurture': *Ann. Eug.* 13, 197–205.

102 HALL, K. R. L., 1965. 'Social organization of old-world monkeys and apes': *Symp. Zool. Soc. Lond.* 14, 265–90.

103 HALL, R., 1909. 'Notes on the magpie (*Gymnorhina leuconota* Gld.)': *Emu* 9, 16–21.

104 HANSSON, A., 1947. 'Physiology of reproduction in mink (*Mustela visor* Schreb) with special reference to delayed implantations': *Acta Zool.* 28, 1–136.

105 HARRIS, G. W., 1955. *Neural control of the pituitary gland:* Edward Arnold, London.

106 HARRIS, G. W., 1959. 'The nervous system – follicular ripening, ovulation, and estrous behavior': in *Recent Progress in the Endocrinology of Reproduction:* C. W. Lloyd, ed., New York Acad. Press, 21–52.

107 HARRIS, G. W., and MICHAEL, R. P., 1964. 'The activation of sexual behaviour by hypothalamic implants of oestrogen': *J. Physiol.* 171, 275–301.

108 HARRIS, G. W., MICHAEL, R. P., and SCOTT, P. P., 1958. 'Neurological site of action of stilboestrol in eliciting sexual behaviour': in *CIBA Foundation Symposium on the Neurological Basis of Behaviour*, Little, Brown & Co., Boston, 236–54.

109 HASKELL, P. T., 1956. 'Hearing in certain Orthoptera. II. The nature of the response of certain receptors to natural and imitation stridulation': *J. exp. Biol.* 33, 767–76.

110 HEDIGER, H., 1950. *Wild Animals in Captivity:* Butterworths, London.

111 HERNANDEZ-PEON, R., SCHERRER, H., and JOUVET, M. 1956. 'Modification of electrical activity in cochlear nucleus during 'attention'' in unanaesthetized cats': *Science, N.Y.* 123, 331–2.

112 HINDE, R. A., 1953. 'The conflict between drives in the courtship and copulation of the chaffinch': *Behaviour* 5, 1–31.

113 HINDE, R. A., 1955. 'A comparative study of the courtship of certain finches (Fringillidae)': *Ibis* 97, 706–45.

114 HINDE, R. A., 1956. 'The behaviour of certain Cardueline F_1 interspecific hybrids': *Behaviour* 9, 202–13.

115 HINDE, R. A., 1956. 'The biological significance of the territories of birds': *Ibis* 98, 340–69.

116 HINDE, R. A., and TINBERGEN, N., 1958. 'The comparative study of species-specific behavior': in *Behavior and Evolution:* A. Roe and G. G. Simpson, eds., New Haven, 251–68.

117 HOBSON, B. M., 1952. 'Routine pregnancy diagnosis and quantitative estimation of chorionic gonadotrophin using female *Xenopus laevis*': *J. Obstet. Gynaec.* 59, 352–62.

118 HOLST, E. VON, and SAINT PAUL, U. V., 1963. 'On the functional organization of drives': *Anim. Behav.* 11, 1–20.

119 HÖRMANN-HECK, S. VON, 1957. 'Untersuchungen über den Erbgang einiger Verhaltensweisen bei Grillenbastarden (*Gryllus campestris* L. × *Gryllus bimaculatus* De Geer)': *Z. Tierpsychol.* 14, 137–83.

120 HOUSE, E., and PANSKY, B., 1960. *A Functional Approach to Neuroanatomy:* McGraw-Hill Book Co., N.Y.

121 HOWARD, H. E., 1920. *Territory in Bird Life:* London.

122 HOWARD, H. E., 1935. 'Territory and food': *Br. Birds* 28, 285–7.

123 HUNSAKER, D., 1962. 'Ethological isolating mechanisms in the *Sceloporus torquatus* group of lizards': *Evolution* 16, 62–74.

123a HUXLEY, J. S., 1914. 'The courtship habits of the great crested grebe (*Podiceps cristatus*)': *Proc. Zool. Soc. London* 1914(2), 491–562.

124 HUXLEY, J. S., 1923. 'Courtship activities in the Red-Throated Diver (*Columbus stellatus* Pontopp); together with a discussion on the evolution of courtship in birds': *J. Linn. Soc. Lond.* 25, 253–92.

125 HUXLEY, J. S., 1938. 'The present standing of the theory of sexual selection': in *Evolution:* G. R. de Beer, ed., Oxford Univ. Press, 11–42.

126 IERSEL, J. J. A. VAN, 1953. 'An analysis of the parental behaviour of the

male three-spined stickleback (*Gasterosteus aculeatus* L.)': *Behaviour Suppl.* 3, 1–159.

127 IERSEL, J. J. A. VAN, and BOL, A. A. C., 1958. 'Preening of two tern species. A study on displacement activities': *Behaviour* 13, 1–88.

128 JAKWAY, J. S., 1959. 'Inheritance of patterns of mating behaviour in the male guinea pig': *Anim. Behav.* 7, 150–62.

129 JASPER, H. H. (ed.), 1958. *The Reticular Formation of the Brain:* Little, Brown & Co., Boston.

130 KENT, G. C., JR, and LIBERMAN, M. J., 1949. 'Induction of psychic estrus in the hamster with progesterone administered via the lateral brain ventricle': *Endocrinology* 45, 29–32.

131 KIMBLE, G. A., and KENDALL, J. W., 1953. 'A comparison of two methods of producing experimental extinction': *J. exp. Psychol.* 45, 87–90.

132 KISLAK, W., and BEACH, F. A., 1955. 'Inhibition of aggressiveness by ovarian hormones': *Endocrinology* 54, 654–92.

133 KLINGHAMMER, E. H., and HESS, E. H., 1964. 'Parental feeding in ring doves (*Streptopelia roseogrisea*): innate or learned?': *Z. Tierpsychol.* 21, 338–47.

134 KLUIJVER, H. N., 1933. *Versl. Plziekt. Dienst. Wageningen* 69, 1.

135 KOCH, H. J., 1942. 'Cause physiologique possible des migrations des animaux aquatiques': *Ann. Soc. Roy. Zool. Belgique* 73, 57–62.

136 KOIVISTO, I., 1965. 'Behavior of the Black Grouse (*Lyrurus tetrix* L.) during the spring display': *Riistatieteellisia Julkaisuja* (Finnish Game Foundation) 26, 1–60.

137 KORTLANDT, A., 1940. 'Wechselswirkung zwischen Instinkten': *Archs néerl. Zool.* 4, 442–520.

138 KUFFLER, S. W., 1953. 'Discharge patterns and functional organization of the mammalian retina': *J. Neurophysiol.* 16, 37–68.

139 LACK, D., 1939. 'The display of the blackcock': *Br. Birds*, 32, 290–303.

140 LACK, D., 1946. 'Blackcock display': *Br. Birds*, 39, 287–8.

141 LACK, D., 1954. *The Natural Regulation of Animal Numbers:* Oxford.

142 LEHRMAN, D. S., 1953. 'A critique of Konrad Lorenz's theory of instinctive behavior': *Q. Rev. Biol.* 28, 337–63.

143 LEHRMAN, D. S., 1955. 'The physiological basis of parental feeding behavior in the Ring Dove (*Streptopelia risoria*)': *Behaviour* 7, 241–86.

144 LEHRMAN, D. S., 1956. 'On the organization of maternal behavior and the problem of instinct': in *L'Instinct dans le comportement des Animaux et de L'Homme:* P. P. Grassé, ed., Masson & Cie., Paris, 475–520.

145 LEHRMAN, D. S., 1958. 'Effect of female sex hormones on incubation behavior in the ring dove (*Streptopelia risoria*)': *J. comp. physiol. Psychol.* 51, 142–5.

146 LEHRMAN, D. S., 1959. 'Hormonal responses to external stimuli in birds': *Ibis* 101, 478–96.

147 LEHRMAN, D. S., 1961. 'Hormonal regulation of parental behavior in

birds and infrahuman mammals': in *Sex and Internal Secretions:* 3rd ed., W. C. Young, ed., Ballière, Tindall & Cox Ltd., Baltimore, 1268–382.

148 LEHRMAN, D. S., 1962. 'Interaction of hormonal and experiential influences on the development of behavior': in *Roots of Behavior:* E. L. Bliss, ed., Harper & Brothers, 142–56.

149 LEHRMAN, D. S., 1964. 'The reproductive behavior of Ring Doves': *Scient. Am.* 211, 48–54.

150 LEHRMAN, D. S., 1964. 'Control of behavior cycles in reproduction': in *Social Behavior and Organization among Vertebrates:* W. Etkin, ed., Univ. of Chicago Press, 143–66.

151 LE MAGNEN, J., 1952. 'Les phenomènes olfactosexuals': *Archs Sci. physiol.* 6, 125–60.

152 LE MAGNEN, J., 1952. 'Les phenomènes olfactosexuals chez le rat blanc': *Archs Sci. physiol.* 6, 295–331.

153 LE MAGNEN, J., 1953. 'L'olfaction: le fonctionnement olfactif et son intervention dans les regulations psycho-physiologiques': *J. Physiol. Path. Gén.* 45, 285–326.

154 LEVINE, L., 1958. 'Studies on sexual selection in mice. I. Reproductive competition between albino and black agouti males': *Am. Nat.* 92, 21–6.

155 LEYHAUSEN, P., 1956. 'Verhaltensstudien an Katzen': *Z. Tierpsychol. Beiheft* 2.

156 LIND, H., 1959. 'The activation of an instinct caused by a "transitional action" ': *Behaviour* 14, 123–35.

157 LOHER, W., 1957. 'Untersuchungen über den Aufban und die Entstehung der Gesänge einiger Feldheuschreckenarten und den Einfluss von Lautzeichen auf das akustische Verhalten': *Z. vergl. Physiol.* 39, 313–56.

158 LORENZ, K., 1931. 'Beiträge zur Ethologie sozialer Corviden': *J. Orn., Lpz.* 79, 67–127.

159 LORENZ, K., 1935. 'Der Kumpan in der Umwelt des Vogels. Der Artgenosse als auslosendes Moment sozialer Verhaltungsweisen': *J. Orn., Lpz.* 83, 137–213, 289–413.

160 LORENZ, K., 1937. 'The companion in the bird's world': *Auk* 54, 245–73.

161 LORENZ, K., 1938. 'A contribution to the comparative sociology of colonial nesting birds': *Proc. 8th Int. orn. Congr., Oxford, 1934* 207–18.

162 LORENZ, K., 1941. 'Vergleichende Bewegungsstudien an Anatinen': *J. Orn., Lpz.* 89, Erg Band 3, 194–294.

163 LORENZ, K., 1951–2. 'Comparative studies on the behaviour of Anatinae': tr. of Lorenz, 1941; tr. by C. H. D. Clarke, *Avicult. Mag.* 57, 157–82; 58, 8–17, 61–72, 86–94, 172–84; 59, 24–34, 80–9.

164 LORENZ, K., 1952. *King Solomon's Ring:* Methuen & Co., London.

165 LORENZ, K., 1953. 'Die Entwicklung der vergleichenden Verhaltensforschung in den letzten 12 Jahren': *Verh. dtsch. Zool. Ges. Freiburg* 1952, 36–58.

166 LORENZ, K., 1958. 'The evolution of behavior': *Scient. Am.* 199, 67–78.

167 Lorenz, K., 1966. *The Nature of Aggression:* Methuen & Co., London.
168 McDermott, F. A., 1917. 'Observations on the light emission of American Lampyridae': *Canad. Ent.* 49, 53–61.
168a Maclean, P. D., 1958. 'Contrasting functions of limbic and neocortical systems of the brain and their relevance to psycho-physiological aspects of medicine': *Am. J. Med.* 25, 615.
169 Magnus, D. B. E., 1950. 'Beobachtungen zur Balz and Eiablage des Kaisermantels *Argynnis paphia* L. (Lep: Nymphalidae)': *Z. Tierpsychol.* 7, 435–49.
170 Magnus, D. B. E., 1958. 'Experimental analysis of some "over-optimal" sign stimuli in the mating behaviour of the fritillary butterfly *Argynnis paphia* L. (Lep. Nymphalidae)': *Proc. 10th Int. Congr. Ent. Montreal* 2, 405–18.
171 Manning, A., 1959. 'The sexual behaviour of two sibling *Drosophila* species': *Behaviour* 15, 123–45.
172 Manning, A., 1961. 'The effects of artificial selection for mating speed in *Drosophila melanogaster*': *Anim. Behav.* 9, 82–92.
173 Manning, A., 1962. 'A sperm factor affecting the receptivity of *Drosophila melanogaster* females': *Nature, Lond.* 194, 252–3.
174 Manning, A., 1965. '*Drosophila* and the evolution of behaviour': in *Viewpoints in Biology* 4 Butterworths, 123–67.
175 Markee, J. E., Sawyer, C. H., and Hollinshead, W. H., 1946. 'Activation of the anterior hypophysis by electrical stimulation in the rabbit': *Endocrinology* 38, 345–57.
176 Marler, P., 1959. 'Developments in the study of animal communication': in *Darwin's Biological Work: Some Aspects Reconsidered:* P. R. Bell, ed., Cambridge Univ. Press, 150–206.
177 Marler, P., 1961. 'The filtering of external stimuli during instinctive behaviour': in *Current Problems in Animal Behaviour:* W. H. Thorpe and O. L. Zangwill, eds., Cambridge Univ. Press, 150–66.
178 Marshall, A. J., 1951. 'The refractory period of testis rhythm in birds and its possible bearing on breeding and migration': *Wilson Bull.* 63, 238–61.
179 Marshall, A. J., 1954. *Bower Birds:* Oxford University Press, London.
180 Marshall, A. J., 1959. 'Internal and environmental control of breeding': *Ibis* 101, 456–78.
181 Marshall, F. H. A., 1936. 'Sexual periodicity and the causes which determine it': *Phil. Trans. R. Soc.* B 226, 423–56.
182 Marshall, F. H. A., 1942. 'Exteroceptive factors in sexual periodicity': *Biol. Rev.* 17, 68–90.
183 Marshall, F. H. A., 1956. 'The breeding season': in *Marshall's Physiology of Reproduction:* 3rd ed., A. S. Parkes, ed., Vol. 1, Longmans, Green & Co. Inc., London, 1–42.
184 Matthews, L. H., 1937. 'The female sexual cycle in the British Horse-

shoe Bats, *Rhinolophus ferrum – Equinum insulanus*, Barrett-Hamilton and *R. hipposideros minutus* Montagu': *Trans. Zool. Soc. London* 23, 224–55.

185 MATTHEWS, L. H., 1939. 'Visual stimulation and ovulation in pigeons': *Proc. R. Soc. Lond.* B 126, 557–60.

186 MAYNARD SMITH, J., 1956. 'Fertility, mating behaviour and sexual selection in *Drosophila subobscura*': *J. Genet.* 54, 261–79.

187 MAYNARD SMITH, J., 1958. 'Sexual selection': in *A Century of Darwin:* S. A. Barnett, ed., Heinemann.

188 MERRELL, D. J., 1949. 'Selective mating in *Drosophila melanogaster*': *Genetics* 34, 370–89.

189 MEYERRIECKS, A. J., 1960. 'Comparative breeding behavior of four species of North American herons': *Publ. Nuttall Orn. Club* 2, 1–158.

190 MOORE, C. R., SIMMONS, G. F., WELLS, L. J., ZALESKY, M., and NELSON, W. O., 1934. 'On the control of reproductive activity in an annual-breeding mammal (*Citellus tridecemlineatus*):' *Anat. Rec.* 60, 279–89.

191 MORRIS, D. J., 1954. 'The reproductive behaviour of the zebra finch (*Poephila guttata*) with special reference to pseudofemale behaviour and displacement activities': *Behaviour* 6, 271–322.

192 MORRIS, D. J., 1954. 'The reproductive behaviour of the river bullhead (*Cottus gobio* L.) with special reference to the fanning activity': *Behaviour* 7, 1–32.

193 MORRIS, D. J., 1955. 'The courtship dance of the swordtail': *Aquarist* March 1955.

194 MORRIS, D. J., 1956. 'The feather postures of birds and the problem of the origin of social signals': *Behaviour* 9, 75–113.

195 MORRIS, D. J., 1956. 'The function and causation of courtship ceremonies': in *L'Instinct dans le comportement des Animaux et de L'Homme:* P. P. Grass, ed., Masson & Cie., Paris, 261–87.

196 MORRIS, D. J., 1957. 'The reproductive behaviour of the bronze mannikin (*Lonchura cucullata*)': *Behaviour* 11, 156–201.

197 MORRIS, D. J., 1957. ' "Typical intensity" and its relationship to the problem of ritualization': *Behaviour* 11, 1–12.

198 MOYNIHAN, M., and HALL, M. F., 1954. 'Hostile, sexual and other social behaviour patterns of the spice finch (*Lonchura punctulata*) in captivity': *Behaviour* 7, 33–76.

199 NICE, M. M., 1941. 'The role of territory in bird life': *Am. Midl. Nat.* 26, 441–87.

200 NOBLE, G. K., 1938. 'Sexual selection among fishes': *Biol. Rev.* 13, 133–58.

201 NOBLE, G. K., and ARONSON, L. R., 1942. 'The sexual behavior of Anura. 1. The normal mating pattern of *Rana pipiens*': *Bull. Am. Mus. nat. Hist.* 80, 127–42.

202 NOBLE, G. K., and CURTIS, B., 1939. 'The social behavior of the

jewel fish *Hemichromis bimaculatus* Gill.': *Bull. Am. Mus. nat. Hist.* 76, 1–46.

203 OLDS, J., 1958. 'Self stimulation of the brain': *Science, N.Y.* 127, 315–25.

204 OLDS, J., 1958. 'Selective effects of drives and drugs on "reward" systems of the brain': in *CIBA Foundation Symposium on the Neurological Basis of Behavior:* Little, Brown & Co., Boston, 124–48.

205 OLDS, J., and MILNER, P., 1954. 'Positive reinforcement produced by electrical stimulation of septal area and other regions of the rat brain': *J. comp. physiol. Psychol.* 47, 419–27.

206 PARKES, A. S., and BRUCE, H. M., 1961. 'Olfactory stimuli in mammalian reproduction': *Science, N.Y.* 134, 1049–54.

207 PATTERSON, I. J., 1965. 'Timing and spacing of broods in the black-headed gull': *Ibis* 107, 433–59.

208 PELKWIJK, J. J. TER, and TINBERGEN, N., 1937. 'Eine reizbiologizche Analyse einiger Verhaltensweisen von *Gasterosteus aculeatus* L.': *Z. Tierpsychol.* 1, 193–204.

209 PERDECK, A. C., 1958. 'The isolating value of specific song patterns in two sibling species of grasshoppers (*Chorthippus brunneus* Thunb. and *C. biguttulus* L.)': *Behaviour* 12, 1–75.

209a PETERS, H. M., 1948. *Grundfragen der Tierpsychologie*: Stuttgart, 23.

210 PETTIT, C., 1959. 'De la nature des stimulations responsables de la sélection sexualle chez *Drosophila melanogaster*': *C.R. Acad. Sci., Paris* 248, 3484–5.

211 PFEIFFER, C. A., 1936. 'Sexual differences of the hypophyses and their determination by the gonads': *Am. J. Anat.* 58, 195–225.

212 PFEIFFER, C. A., 1937. 'Hypophyseal gonadotropic hormones and the luteinization phenomenon in the rat': *Anat. Rec.* 67, 158-75.

213 POLIKARPOVA, E., 1940. 'Influence of external factors upon the development of the sexual gland of the sparrow': *C.R. Acad. Sci. U.R.S.S.* 26, 91–5.

214 PORTER, R. W., CAVANAUGH, E. B., CRITCHLOW, B. V., and SAWYER, C. H., 1957. 'Localized changes in electrical activity of the hypothalamus in estrous cats following vaginal stimulation': *Am. J. Physiol.* 189, 145–51.

215 PORTMAN, A., 1961. *Animals as Social Beings*: Hutchison.

216 PRESCOTT, R. G. W., 1966. 'Estrous cycle in the rat: effects on self-stimulation behaviour': *Science, N.Y.* 152, 796–7.

217 PRINGLE, J. W. S., 1954. 'A physiological analysis of cicada song': *J. exp. Biol.* 31, 525–60.

218 PYCRAFT, W. P., 1914. *The Courtship of Animals:* Hutchison.

219 QUINLAN, J., BISSCHOP, J. H. R., and ADELAAR, T. F., 1941. 'Bionomic studies on cattle in the semi-arid regions of the Union of South Africa. IV. The ovarian cycle of heifers during summer': *Onder-*

stepoort J. Vet. Sci. Anim. Indust. 16, 213–41. (Taken from: *Biol. Absts.* 18, 4714 (1944).)

220 RASMUSSEN, A. T., 1921. 'The hypophysis cerebri of the woodchuck (*Marmota monax*) with special reference to hibernation and inanition': *Endocrinology* 5, 33–66.

221 RASMUSSEN, E. W., 1952. 'The relation between strength of sexual drive and fertility as evident from experimental investigation': *Proc. 2nd Int. Congr. Anim. Repro. (Copenhagen)* 1, 188–91.

222 REED, S. C., and REED, E. W., 1950. 'Natural selection in laboratory populations of *Drosophila*. II. Competition between white-eye gene and its wild type allele': *Evolution* 4, 34–42.

223 RENDEL, J. M., 1945. 'Genetics and karyology of *Drosophila subobscura*. II. Normal and selective matings in *Drosophila subobscura*': *J. Genet.* 46, 287–302.

224 RENDEL, J. M., 1951. 'Mating of ebony, vestigial and wild type *Drosophila melanogaster* in light and dark': *Evolution* 5, 226–30.

225 RICE, V. A., 1942. *Breeding and Improvement in Farm Animals:* 3rd ed., McGraw-Hill, N.Y.

226 RICHARDS, O. W., 1927. 'Sexual selection and allied problems in the insects': *Biol. Rev.* 2, 298–364.

227 RICHTER, C. P., 1927. 'Animal behavior and internal drives': *Q. Rev. Biol.* 2, 307–43.

228 RISS, W., VALENSTEIN, E. S., SINK, J., and YOUNG, W. C., 1955. 'Development of sexual behavior in male guinea pigs from genetically different stocks under controlled conditions of androgen treatment and ageing': *Endocrinology* 57, 139–46.

229 ROEDER, K. D., 1962. 'The behaviour of free flying moths in the presence of artificial ultrasonic pulses': *Anim. Behav.* 10, 300–4.

230 ROEDER, K. D., and TREAT, A. E., 1961. 'The detection and evasion of bats by moths': *Am. Sc.* 49, 135–48.

231 ROMER, A. S., 1950. *The Vertebrate Body:* W. B. Saunders & Co., Philadelphia and London.

232 ROOT, W. S., and BARD, P., 1937. 'Erection in the cat following removal of the lumbo-sacral segments': *Am. J. Physiol.* 119, 392.

233 ROSENBLATT, J. S., and ARONSON, L. R., 1958. 'The decline of sexual behavior in male cats after castration with special reference to the role of prior sexual experience': *Behaviour* 12, 285–338.

234 ROTH, L. M., 1948. 'An experimental laboratory study of the sexual behaviour of *Aedes aegypti* (1)': *Am. Midl. Nat.* 40, 265–352.

235 ROTH, L. M., and WILLIS, E. R., 1952. 'A study of cockroach behaviour': *Am. Midl. Nat.* 47, 66–129.

236 ROTH, L. M., and WILLIS, E. R., 1954. 'The reproduction of cockroaches': *Smiths. Misc. Coll.* 122, 1–49.

237 ROTHENBUHLER, W. C., 1964. 'Behaviour genetics of nest cleaning in

honey bees. IV. Responses of F_1 and backcross generations to disease-killed brood': *Am. Zoologist* 4, 111–23.

238 ROTHENBUHLER, W. C., 1964. 'Behaviour genetics of nest cleaning in honey bees. I. Responses of four inbred lines to disease-killed brood': *Anim. Behav.* 12, 578–83.

239 ROWAN, W., 1938. 'Light and seasonal reproduction in animals': *Biol. Rev.* 13, 374–402.

240 ROWELL, C. H. F., 1961. 'Displacement grooming in the chaffinch': *Anim. Behav.* 8, 38–63.

241 RUSSELL, W. M. S., 1954. 'Experimental studies of the reproductive behaviour of *Xenopus laevis*. I. The control mechanisms for clasping and unclasping, and the specificity of hormone action': *Behaviour* 7, 113–88.

242 SAWYER, C. H., 1959. 'Nervous control of ovulation': in *Recent Progress in the Endocrinology of Reproduction:* C. W. Lloyd, ed., N.Y. Acad. Press Inc., 1–20.

243 SAWYER, C. H., and KAWAKAMI, M., 1959. 'Characteristics of behavioral and electroencephalographic after-reactions to copulation and vaginal stimulation in the female rabbit': *Endocrinology* 65, 622–30.

244 SAWYER, C. H., and ROBINSON, B., 1956. 'Separate hypothalamic areas controlling pituitary gonadotroppic function and mating behaviour in female cats and rabbits': *J. clin. Endocrin. Metabolism* 16, 914–15.

245 SCHENKEL, R., 1947. 'Ausdrucks-Studien an Wölfen': *Behaviour* 1, 81–130.

246 SCHNEIDER, D., 1957. 'Elektrophysiologische Untersuchungen von Chemo- und Mechanorezeptoren der Antenne des Seidenspinners *Bombyx mori* L.': *Z. vergl. Physiol.* 40, 8–41.

247 SELOUS, E., 1909–10. 'An observational diary on the nuptial habits of the black cock (*Tetrao tetrix*) in Scandinavia and England': *Zoologist* 13, 400–13; 14, 23–9, 51–6, 176–82, 248–65.

248 SEVENSTER, P., 1961. 'A causal analysis of a displacement activity (fanning in *Gasterostens aculeatus* L.)': *Behaviour Suppl.* 9, 1–170.

249 SHERRINGTON, C. S., 1906. *The Integrative Action of the Nervous System:* New York.

250 SIMMONS, K. E. L., 1955. 'The interpretation of the house-sparrow display': *Ibis* 97, 159–60.

251 SMITH, O. R., 1935. 'The breeding habits of the Stone Roller Minnow (*Campostoma anomalum* Rafinesque)': *Trans. Am. Fish Soc.* 65, 148–51.

252 SNOW, D. W., 1958. *A Study of Blackbirds:* Allen & Unwin.

253 SPIETH, H. T., 1952. 'Mating behavior within the genus *Drosophila* (Diptera)': *Bull. Am. Mus. nat. Hist.* 99, 401–74.

254 STEIN, R. C., 1956. 'A comparative study of "advertising song" in the *Hylocichla* thrushes': *Auk* 73, 503–12.

255 STONE, C. P., 1923. 'Experimental studies of two important factors

underlying masculine sexual behavior. The nervous system and the internal secretion of the testis': *J. exp. Psychol.* 6, 85–106.

256 STONE, C. P., 1932. 'The retention of copulatory ability in male rabbits following castration': *J. genet. Psychol.* 40, 296–305.

257 STONE, C. P., 1939. 'Copulatory activity in adult male rats following castration and injections of testosterone propionate': *Endocrinology* 24, 165–74.

258 STONE, C. P., 1939. 'Sex drive': in *Sex and Internal Secretions:* 2nd ed., W. C. Young, ed., Williams & Wilkins, Baltimore.

258a STONOR, C. R., 1940. *Courtship and Display among Birds:* Country Life, 38.

259 STRIDE, G., 1956. 'On the courtship behaviour of *Hypolimnas misippus* L. (Lepidoptera: Nymphalidae), with notes on the mimetic association with *Danaus chrysippus* L. (Lepidoptera: Danaidae)': *Br. J. Anim. Behav.* 4, 52–68.

260 STUBBS, F. J., 1910. 'Ceremonial gatherings of the magpie': *Br. Birds* 3, 334–6.

261 SUMMERS-SMITH, D., 1954. 'The communal display of the house-sparrow': *Ibis* 96, 116–28.

262 SUMMERS-SMITH, D., 1954. 'Colonial behaviour in the house-sparrow': *Br. Birds* 47, 249–65.

263 SUMMERS-SMITH, D., 1955. 'Display of the house-sparrow (*Passer domesticus*)': *Ibis* 97, 296–305.

264 SUMMERS-SMITH, J. D., 1963. *The House Sparrow:* Collins.

265 THORPE, W. H., 1963. *Learning and Instinct in Animals:* 2nd ed., Methuen & Co., London.

266 TINBERGEN, N., 1939. 'On the analysis of social organization among vertebrates with special reference to birds': *Am. Midl. Nat.* 21, 210–34.

267 TINBERGEN, N., 1939. 'Field observations of East Greenland birds. II. The behavior of the Snow Bunting (*Plectrophenax nivalis subnivalis* A. E. Brahm) in spring': *Trans. Linn. Soc. N.Y.* 5, 1–94.

268 TINBERGEN, N., 1942. 'An objectivistic study of the innate behaviour of animals': *Biblioth. biotheor.* 1, 39–98.

269 TINBERGEN, N., 1951. *The Study of Instinct:* Clarendon Press, Oxford.

270 TINBERGEN, N., 1952. '"Derived activities", their causation, biological significance, origin and emancipation during evolution': *Q. Rev. Biol.* 27, 1–32.

271 TINBERGEN, N., 1954. 'The origin and evolution of courtship and threat display': in *Evolution as a Process:* J. S. Huxley, ed., Allen & Unwin, London, 233–251.

272 TINBERGEN, N., 1957. 'The function of territory': *Bird Study* 4, 14–27.

273 TINBERGEN, N., 1959. 'Comparative studies of the behaviour of gulls (Laridae): a progress report': *Behaviour* 15, 1–70.

274 TINBERGEN, N., and BROEKHUYSEN, G. J., 1954. 'On the threat and

courtship behaviour of Hartlaub's Gul (*Hydrocoloeus novae-hollandiae hartlaubi* (Bruch))': *Ostrich* 25, 50–61.

275 TINBERGEN, N., and MOYNIHAN, M., 1952. 'Head flagging in the black-headed gull; its function and origin': *Br. Birds* 45, 19.

276 VALENSTEIN, E. S., and GOY, R. W., 1957. 'Further studies of the organization and display of sexual behavior in male guinea pigs': *J. comp. physiol. Psychol.* 50, 115–19.

277 VALENSTEIN, E. S., RISS, W., and YOUNG, W. C., 1955. 'Experiential and genetic factors in the organization of sexual behavior in male guinea pigs': *J. comp. physiol. Psychol.* 48, 397–403.

278 VALENSTEIN, E. S., and YOUNG, W. C., 1955. 'An experiential factor influencing the effectiveness of testosterone propionate in eliciting sexual behavior in male guinea pigs': *Endocrinology* 56, 173–7.

279 VERWEY, J., 1930. 'Die Paarungsbiologie des Fischreihers': *Zool. Jb. Physiol.* 48, 1–120.

280 WALSH, E. G., 1964. *Physiology of the Nervous System:* 2nd ed., Longmans, Green, London.

281 WEBER, H., 1924. 'Liebesspiele, Eiübertragung, und Copulation bei *Hippocampus brevirostris*': *Zool. Anz.* 60, 281–90.

282 WEIDMANN, U., 1955. 'Some reproductive activities of the common gull, *Larus canus*': *Ardea* 43, 85–132.

283 WEIH, A. S., 1951. 'Untersuchungen über das Wechselsingen (Anaphonie) und über das angeborene Lautschema einiger Feldheuschrecken': *Z. Tierpsychol.* 8, 1–41.

284 WHEELER, R. S., 1943. 'Normal development of the pituitary in the opossum and its responses to hormonal treatments': *J. Morph.* 73, 43–87.

285 WHITTEN, W. K., 1957. 'Effect of exteroceptive factors in the oestrous cycle of mice': *Nature, Land.* 180, 1436.

286 WHITTEN, W. K., 1959. 'Occurrence of anoestrus in mice caged in groups': *J. Endocr.* 18, 102.

287 WIGGLESWORTH, V. B., 1949. 'The light of glow-worms and fireflies': *Science News* 12, 9–22.

288 WIGGLESWORTH, V. B., 1953. *The Principles of Insect Physiology:* Methuen, London.

289 WITHERBY, H. F., JOURDAIN, F. C., TICEHURST, N. F., and TUCKER, B. W., 1945. *The Handbook of British Birds:* Witherby, London.

290 WOLFE, J. M., CLEVELAND, R., and CAMPBELL, M., 1932. 'Cyclic histological changes in the anterior hypophysis of the dog': *Anat. Rec.* 52, Suppl. Feb. 1932, Abstract 98, 44.

291 WOOD-GUSH, D. G. M., 1954, 'The courtship of the Brown Leghorn cock': *Br. J. Anim. Behav.* 2, 95–102.

292 WOOD-GUSH, D. G. M., 1956. 'The agonistic and courtship behaviour of the Brown Leghorn cock': *Br. J. Anim. Behav.* 4, 133–42.

293 WOOD-GUSH, D. G. M., 1960. 'A study of sex drive of two strains of cockerel through three generations': *Anim. Behav.* 8, 43–53.
294 WOOD-GUSH, D. G. M., and OSBORNE, R., 1956. 'A study of differences in the sex drives of cockerels': *Br. J. Anim. Behav.* 4, 102–10.
295 YEATES, G. K., 1934. *The Life of the Rook:* London.
296 YOUNG, W. C., 1957. 'Genetic and psychological determinants of sexual behavior patterns': in *Hormones, Brain Function and Behavior:* M. B. Hoagland, ed., N.Y. Acad. Press, 75–98.
297 YOUNG, W. C., 1961. 'The hormones and mating behaviour': in *Sex and Internal Secretion.* 3rd ed., W. C. Young, ed., Baillière, Tindall, & Cox Ltd.

Index

Index